Animal Husbandry: Top Most Practical Guide on Laudable Techniques of Making Money through Animal Science

HENRY OJO

Copyright © 2017 Henry Ojo

All rights reserved.

ISBN-10: 1979403260
ISBN-13: 978-1979403269

DEDICATION

I dedicate this book to the Almighty God ,And my beloved wife Lady Evangelist Ruth Ojo for her passion for soul wining ,and commitment to God's work and our family, also to all God's children, particularly those of you who endure my ministry in the time when I just began.

CONTENTS

Acknowledgments	i
Introduction	1
Chapter One: **Cattle Production**	3
Chapter Two: **Sheep Production**	19
Chapter Three: **Goats Production**	35
Chapter Four: **Pig Production**	53
Chapter Five: **Rabbit Production in the Tropics**	67
Chapter Six: **Grasscutter**	97
Chapter Seven: **Poultry (Exotic Birds Production)**	107
Chapter Eight: **Poultry (Domestic Ducks Production)**	133
Chapter Nine: **Poultry (Production of Local Chicken or Indigenous Birds)**	143
Chapter Ten: **Snail Production**	149
Chapter Eleven: **Agro-Feasibility Study**	170
Bibliography	180

ACKNOWLEDGMENTS

In God own processes of making His men, He always use various vessels to develop them. On this note, I want to use this medium to acknowledge all men and women whom God has used to contribute and add values to my life.

INTRODUCTION

Livestock production plays a very important role in agricultural economy.

The capital value of the industry is estimated at over N100 Billion and contributes over 5% of the gross Domestic Product (G.D.P).

It provides the high quality animal protein needed for a balance diet and healthy populace, raw materials for the related agro- based industries and plays a significant role in promoting crop production through the use of animal traction and manure for soil fertility.

Livestock production in Nigeria as in many developing countries is not efficient as in developed countries. For example, Africa, Asia and Latin America have over 50% of the world's dairy cows, poultry and pigs but provide only 15% of the world's milk and 30% of the world's eggs and pork.

Nigeria's livestock resources consist of aboLit 13.9

Million cattle, 34.5

Million goats and sheep, 3.5 Million pigs, and about 150 Million poultry

(local and exotic).

The off-take from these resources barely provides less than 50% of the FAO recommended target consumption of 35 grams/caput/day of animal protein which is considered necessary for overall health. Generally, Agriculture in Nigeria has fallen form self-sufficiency status to net importer since the advent of the oil industry. Past and present government of the country realize the importance of Agriculture in any economy whether developed or developing ones. To this extent, quite a number of policies have been formulated to boost agricultural production in the country but these have met with little success.

However, there is considerable scope for achieving self sufficiency in the production of poultry, eggs and meat as well as by- products but this is constrained by the feeds and nutritional problems.

Chapter One

CATTLE PRODUCTION

ORIGIN

The family of animal that includes all types of domestic cattle are known as the Bovidae. They are ruminants with hollow horns and hoofs and an even number of toes. They are among the earlier of all animals to be domesticated for agricultural purpose in Western Europe. There is no evidence of domestication in Paleolithic times, but there are plentiful remain in the Swiss lake dwelling and other deposits of Neolithic age.

All domestic cattle originated from Bos taurus or humpless cattle and Bos indicus (zebu) or humped cattle whose wild ancestors are extinct, and the wild cattle of South Asia or from the crosses of two or three types. They are all inter-fertile to a greater or lesser degree.

ECONOMIC IMPORTANCE OF CATTLE

The economic importance of cattle cannot be over emphasized. Cattle can be used for production of milk as well as beef. Beef cattle should have a large amount of flesh anc:l short legs. Cattle could also be used as work animals. Work cattle or work oxen may be bulls or bullock castrated bulls or steer). The latter are more suitable for work because they are quieter and easier to control.

BREEDS OF CATTLE

All known modern domesticated cattle are usually believed to belong to Bos taurus (European breed such as short horns, Jersey etc); Bos indicus (zebu breed such as Brahman, Harriana etc) or crosses of these two species as American Brahman (Brahman, Santa Gertrudes and Brangus (Brahman and Angus). Breeds as they are known today may not only exist as of resent origin. The definition of a breed is difficult and inexplicit although the term is commonly used and the practice well understood. It may be said generally to connote animal, which for a long time have been selectively bred so that they possess identity in color, size, conformation and function; and these other distinguishing characteristics. This is accompanied by working on the principles of "like begets like". It is only within relatively recent times that the science of genetics has contributed to animal breeding. There are many old established breeds in Europe e.g. the Charolais and the Dutch Freisian of the Netherlands, Campagna di Roma of Italy and many others, but the

British Breeds are of particular interest because of their influence in building up vast herds which furnishes the supplies of breeds and. which other Countries are largely dependent.

THE BEEFBREED

ANGUS:

This is a black polled beef cattle for many years known as Aberdeen Angus. It originated in the country of Aberdeen Scotland.

The characteristic features are black color, polled head, compact and low set body, fine quality of flesh and high dressing percentage. The Angus is a big breed of the highest rank and for years pure cross Angus steers have held high places of fat stock shows in Britain and USA. This breed was introduced into the USA in 1873 and after that date its influence spread widely there and in other countries.

BRAHMAN:

The term Brahman was used by the USA Department of Agriculture as the name of breeds of Indian Cattle in

USA. In South America and Europe these cattle are known as Zebus. Braham Cattle are characterized by prominent hump above the shoulders and an extreme development of loose pendulous skin under the throat and the dew-lap. The pendulous skin provides a greater surface and gives better heat regulation which is important in hot climate. The head is long and narrow ears are long and the horns differ widely according to sex and strain. The color varies form shades of gray to black . Indian cattle like those of Europe vary in size, form and symmetry under the influence of local differences in climate, soil and available feed. In their native homes, Indian cattle are used primarily for work and milk production.

BRANGUS

The name is derived fm parental bretd (Brahman and Angus). The breed had its origin at Clear Creek Ranch in Oklahoma in 1942. The breeding programme was similar to that used by the King Ranch to develop the Santa Gertrudis breed. The foundation stock contained about 5/8 Angus and 3/8 Brahman ancestors.

The Brangus is usually black in color and tends to resemble the Angus in conformation. It has a nearly smooth top line and less dew-lap and less loose skin above the navel than does the Brahman. Although total number of Brangus cattle have been relatively small, the breed proved

itselfuseful in crossing with the native cattle of the coastal region. Like the Angus the breed is polled.

SANTA GERTRUDIS

This breed was developed by the King Ranch in Texas. It originally resulted from crossing Brahman bulls of about 7/8 pure breed and pure breed short horn calves. Over a period of years beginning with the first process in 1910 selective breeding was practiced in which preference was given to red color without sacrificing type and conformation. The cattle are usually solid red in color with ocóasional small white markings on the forehead or in the region of the flank. They were developed from foundation stock of 5/8 shorthorn and 3/8 Brahman ancestry. They have proved to be highly adaptable to the semi tropical gulf coast.

SHORT HORN

This is an example of beef cattle by selection within a breed. In the last quarter of 18th century two brothers Charles and Robert Coiling farming in Durham county England began to improve the local cattle of the Teeswater District. Their efforts were supplemented by other breeders notably from Thomas Bates and Thomas Booth in Yorkshire. As many cattle of this breed that has been exposed to other countries from

Durham are often called by that name. They are found in practically every country in the world. They are numerous in North and south America particularly in Argentina, !n Europe and Australia and have been bred quite extensively in Africa. Strains of Shorthorn has been selected for milk and butter fat production as well as for beef. In USA they are called milking Shorthorns. In Canada they are called dual purpose Shorthorns while in England and Australia Dairy Shorthorns.

In England many herds of Shorthorn cattle still show the beef and milk combinations developed by the early improvers of the breed. The polled shorthorns as the name implies, is a strain within the breed possessing all of the Shorthorn characteristics except horns. This strain was developed in the USA in the late 1880's and early 1890's through the use of naturally hornless registered Shorthorns found within the breed.

AFRICKANDER

This breed originated from the cattle owned by the Hottentots in the 17th and 18th centuries, but some Portuguese workers considered that it originated from an importation of cattle of the Alantejo breed frjn Portugal. Dutch settlers in South Africa acquired this breed as earlyIn 1652. Climatic environment might be described as modified sub-tropical to tropical in the North and Mediterranean in the Cape Province.

They are large, muscular and hardy. The coat color is usually a shade of red, often with white marks on the underline. The hair is short and glossy, the skin thick, pliable and amber to yellow color. The horns are oval in cross-section and have a characteristic twist.

These cattle were originally developed as drought animal, but more recently have been developed for beef production. They are found in many African countries USA, the Philippines and Australia.

Other exotic breeds are Charolal , Hereford and Beefmaster amongst others. Our indigenous beef cattle are West African Dwarf Shorthorn (Muturu), N'dama, Rahaji and Adamawa.

DAIRY BREEDS

THE HOLSTEIN FRIESIAN

This breed originated in North Holland and Freiseland. Its Chief characteristics are large size and black and white spotted markings, sharply defined rather than blended. These cattle are believed to be selected for dairy qualities for about 2000 years. They have long been widely distributed over the more fertile lowland of continental Europe where they are valued highly for their milk producing ability. In the USA the Holstein Friesian outnumber all other diary breeds and produce more than 1/2 of milk supply. In the United Kingdom an average yield is 4,500 litres with a butter fat content of about 3.5 percent

JERSEY

This is believed to have descended from French cattle. Colour is fawn or cream but darker shades are found. The Jersey, relatively small is adaptable to a wide range of condition and distribution worldwide. The milk is remarkably rich in butter fat and for that reason animals of that breed are in demand for mating with native stock to improve butterfat percentage in milk.

Average milk yield is 3,500 litres in the United Kingdom and 2,300 litres in West Africa. The butterfat content of milk is about 5%. The animals are mc're temperamental than any other imported breed.

GUERNSEY

The Channel sIand is the home of Guernsey. Like the Jersey the breed is thought to have descended form France. They are fown-coloured, marked with white and are longer than Jersey breeds. Guernsey are noted for the production of milk of a pronounced yellow colour. They are not good beef cattle.

SWISS BROWN:

This is a native of Switzerland and probably one of the oldest in existence. While this is classified as dairy breed in USA they are often considered dual purpose breed elsewhere as they are heavier-boned and thicker - fleshed than the cattle of the other dairy breeds. A medium sized breeds producing an average of 3,800 liters of milk in its home Country will only provide 2,400 litres in west Africa. The butter fat con*en±is 4 per cent.

AYRSHIER

This originated from Ayr County in South West Scotland in the later part of 18th Century and considered the only special dairy breeds to have originated in the British Isles. The body color from almost pale white to all cherry red or brown with any combination of these colors. The beef quality of the breed is of secondary importance. The distribution is wide and exported to many countries.

Some Nigeria indigenous dairy breeds are the Kuri, SokotoGudali and the Bunaji (white Fulani).

WORK CATTLE

For centuries work animals have been used for transportation and cultivation purpose in all tropical regions of Asia and they were initially used by European Colonists in the American and Australia tropics. In tropical Africa they were used only in those places where they were introduced by the Europeans.

Some physical specification for work animals are compactness, sturdy body with well developed muscles, particularly on the back and the hiquarters, a broad deep chest; strong leg; and large, sound, hard feet. In making a choice, consideration should first be given to a work test or by conformation.

Some select work cattle are Harriana buttocks, droughmaster and good number of beef breeds. Some good indigenous work animals are the Bunaji and SokotoGudali, Muturu and Rahaji.

CATTLE PRODUCTION MANAGEMENT

The management system adopted by any production enterprise depends squarely on the scope of the enterprise, the environment, the objective as well as the market possibilities.

EXTENSIVE SYSTEM (BEEF PRODUCTION)

Most of the beef produced in the tropic comes from animals managed on free range, grazing natural forage that grows on land which is not in use. The extensive beef producers are from those who have small herd of cattle adjacent to their crop farm or nomads or transhumance who migrate with their cattle or ranchers who manage herds on extensive but enclosed areas of herd. In addition to using range grazing, the extensive producers also use crop residues which come from their farms and sometimes from other farmers.

Other sources of beef produced in the tropics, come from work bulls, from peasant intensive producer, from specialist milk producer who raise surplus male calves, under ley farming (alley cropping) conditions or purchased cattle fattened on feed lots.

MIGRATCISHIFTING CULTIVATION

This groupfarmers raise small stock and poultry but cattle herding is carried ouThis system in parts of West Africa and rain forest belt of South America

SEDENTARY SHIFTING CULTIVATION

Where land resources are lean and population is increasing rapidly and opportunities abound for cash crop production, sedentary has replaced migratory shifting cultivation. Typical location are West Africa, South East Asia, South America and North India.

SEDENTARY SUBSISTENCE (Rain-fed or irrigated cultivation)

This is permanent cultivation by subsistence farmers due to production pressure or other factors can force abandonment of shifting cultivation or nomadism. Cattle and buffaloes are the replacements. The locations are Northern Nigeria, Senegal, Uganda, South East Asia, Western Asia and Africa, Caribbean, Indian

Ocean and South America.

NOMADIC HERDING

This is the roaming from place to place for pasture. This practice is today limited in scope.

TRANSHUMANCE OR SEMI-NOMADIC HERDING

This is the seasonal moving of livestock to other regions. It is of very

considerable importance throughout West, North . and East Africa

This system is also practiced in Malagasy Republic, Western Asia and in India subcontinent.

Transhuman. have a village base where they cultivate crops during wet season but move towards the part of the region during dry seasons. Due to their settled life they can be provided with some social services such as education and health.

COMMERCIAL RANCHING

Commercial ranching under extensive system of beef cattle production is the alternative to various types of nomadism. Typical examples found in the tropics are the humid regions of Mexico, Central America, the Caribean and South America. Also some are found on tsetse-free areas of East and South Central Africa among other area.

The major technical objectives of ranch management are to decrease annual fluctuation in cattle numbers and seasonal fluctuations in live weight, maximize reproductive performance and minimize mortality. The breed to be used under this system should be breeds that produce the maximum quality of High-Quality meat in most economic manner.

DISEASE AND PARASITES UNDER EXTENSIVE MANAGERIAL SYSTEM.

It could be correct to say that epidemic diseases and some external parasites will continue to be difficult to control under extensive beef cattle production system. However, with well planned eradication programs, improvement in vaccine production and distribution and improved nutrition and management, disease and parasite will cease to be a major constraint on beef cattle productivity.

MANAGEMENT AND FEEDING UNDER EXTENSIVE SYSTEM

In the management of beef cattle enterprise the integration of all the known improved management practice should be applied for the overall best results.

BREEDING AND REPRODUCTION

For productivity to be maximized under extensive system of beef cattle production each breeding cow and heifer should produce one calf per year and this calf should live. Such factor that can influence calving percentacje under extensive husbandry condition in the tropics such as the utilization of a suitably acclimatized type of cattle, good nutritive regime for breeding females, organised mating, excellent fertility of bulls and appropriate bull to breeding female ratio. (A range of 3-6 percent bulls in the herd should be sufficient).

Poor nutrition can have a marked effect on calving rates in drier areas of Australia, the Americas and Africa. Low total nutrients intake or a deficiency of protein and or minerals may so reduce the calving percentage that the cattle may calve only every 2 years or less often. Supplementary rations is a very effective way of improving calving percentage.

INTENSIVE SYSTEMS

SEDENTARY SUBSISTENCE RAIN- FE /OR/IRRIGATED CULTIVATION.

This system not only intensifies cropping system and increased demand for work animals, it also introduces the possibility of providing forage on a year round basis. Though irrigated land is too expensive to use for forage production it is nevertheless practiced in the tropics.

LEY FARMING

This is the alternate growing of crops and grass for the purpose of producing enough fodder for the animal. Examples are the Kano plain in Nigeria and Teso District in Uganda.

PERENNIAL CROP CULTIVATION

Cattle production could be combined with perennial field crops such as sugar cane, pineapple and sisal or with perennial tree crops such as coconuts, oil palm, rubber and pear. The by-products of perennial field crops are of major importance in cattle feeding. Sugar cane alone for instance provides three feed by-products, green top, molasses and bagasse.

SPECIALIZED CATTLE PRODUCTION SYSTEMS

This system of production is very limited in the tropics. The three major types of operation are:

1. large specialized milk producers, who grow out and fatten surplus m&e caves

2. specialized beef producers using intensively managed planted pasture and or planted pasture trees

3. specialized beef producers purchasing cattle from other producers and growing them out and or fattening then,in feedlots.

WORK CATTLE

For centuries work animals have been used for transport and cultivation purpose in all tropical regions of Asia and they were initially used by European colonist in the America and Australian tropics. However the introduction of powered vehicles is very rapidly removing the burden of road transport from animal throughout the world.

Chapter Two

SHEEP PRODUCTION

ORIGIN

Sheep are classified in the sub-family caprinae and all domestic sheep are included in the genus Ovis aries. There are four species:

1. Ovis Musimon (the moufflon)

2. Ovis orientalis (the Urial)

3. Ovis ammon (the Agali)

4. Ovis canadensis (the Bighorn)

According to Zeuner (1963) sheep were domesticated at about the same time as goats and before crop agriculture. He considered original centre of domestication to be Aralo-Caspian steppe. From here sheep breeding spread in to South Asia, to Western Asia and on into Europe and Africa.

BREEDS

WEST AFRICAN DWARF

This breed is common all over the humid area of West Africa. The coat is usually white, mixture of black and white or brown. The rams are horned. Average mature weight is approximately 36kg.

THE OUDA (WEST AFRICA LONG LEGGED OR FULANI)

The breed represents a type of sheep found throughout the Sahel and Savana Zone of tropical Africa from Western Ethiopia to Northern Nigeria. Mature males measure up to 84cm at the shoulder. They are both hairy and woolled sheep. Mature males and females weigh an average 68kg and 41kg respectively. Horns are quite large and curve downward. The color is solid-black or brown.

YAN KASSA

It has short hair and found mainly in Northern Nigeria. White in color and have black extremities. The ewe, (female) may be polled (without horn) or very short horns.

SUDANESE DESERT

This is a large, fat-tailed, long-legged, hairy breed. The

coat color is normally brown or pied. They p ess long lop ears, a long tapering tail which sometimes carries much fat. Mature rams weigh up to 68kg.

11

NILOTIC

Small, woolled, and short-legged, thin-tailed breed. The fleece is short, fine and usually white in color, although may possess black and tan patches. Mature weight is approximately 11 kg. The ewes are poor milkers and adapted to swampy environment.

1-IAGAWA

This is a small, polled, fat-tail breed owned by a transhuman tribe in the Western Sudan. It has long black coat.

MASAI

They are large, long-legged, coarse- woolled sheep owned by the Masai, who live in Kenya and Tanzania. They are representative of a large number of different breeds of fat-tailed sheep found throughout Northeast, East and central Africa

SOMALI

This is a fat-rumped Somali breed closely resembles the Black head Persian. The limbs and under parts are wholly white. Mature males weigh up to 42kg. Has excellent mutton and very valuable skin.

BLACKHEAD PERSIAN

Large, fat-rumped breed, was developed in south Africa from fat rumped

Somali-type sheepfirst imported into Port Beaufort, South Africa.

BLACK BELLIED BPRBADOS

Moderately large, long-legged, hairy breed developed on the Island of Barbados by crossing West African Dwarf sheep, originally imported at the same time as the African slaves who worked on the sugar plantation, with temperate-typed sheep breeds. Coat is usually smooth though rams possess a ruff and is brown to brownish black in color. Males weigh 45kg at one year of age.

CRIOLLO

This breed originally developed in the North of South Africa from the

Churro and the Spanish Merino, has been upgraded. Other African

Breeds include Morado Nova (Brazil), and Brahaman.

HEJAZI

This is a fat-rumped, hairy breed. It typifies the type of sheep common throughout Arabia and around the Horn of Africa that is capable of thriving under arid tropical condition. They possess defined throat ruff. Mature animals weigh up to 32kg.

12

NELLORE

This breed typifies the hairy sheep of South India. They are longlegged, measuring some 76cm but possess quite, broad bodies. The rams are horned and heavily manned.

LOHI

This breed typifies the best mutton and milk breed of India and is Indigenous to the western Punjab. The head is heavy with large lop ears and a short thick neck. Height at shoulders is 79cm. Mature males weigh 68kg.

SHEEP BREEDING

Sheep as well as goats are a seasonal .i.e. they can breed throughout the whole year. From practical observation and experimental evidence available, it can be concluded that it is possible to obtain two crops of lambs each year and quite definitely it might be possible to obtain two crops of lamb each year or three crops in two years as well as in goats. Single, twining, triplet, and quadruplet in sheep are possibilities. The greater the number of litter, the better to select for breeding. When they are pregnant towards the end of gestation increase the level of feeding. This is referred to as "steaming up".

It is possible for a ram to sire (mate) up to 60 ewes in one breeding season though in temperate zone the normal practice is to mate one ram with up to 40 ewes.

DISEASES AND PARASITES

Like all other animals in the tropic sheep also suffer diseases as anthrax, blackleg, navel ill, pulpy kidney, and leptospirosis, though undiagnosed. It is advisable that sanitation, proper feeding regime, and good management should be maintained. The following are some of the tropical diseases of importance:

a. Blue Tongue: A virus disease of sheep that is occasionally found in cattle and wild fauna and is usually transmitted by sand flies of the genusculcoides. Annual vaccination is known to have helped.

b. Rift valley fever: The causative virus is closely related to the yellow fever virus and is only found in Africa, but now in Chad and the Cameron Republic. The vector is mosquito. The vaccine is available.

c. Nairobi sheep disease: this is a tick-borne virus of sheep and goats found in east and south Africa and characterized by an acute haemorhegic septicemia. The vector is Rhipecephalus

appendiculatus. No known treatment or vaccine.

d. Sheep pox: this is a virus disease transmitted by direct contact, probably by droplet inhalation. There is a vaccine that confers one year immunity.

HEART WATER

A disease caused by a rickettsia-type organism that is transmitted by the tick Amblyomma herbarium. Control is through the control of the tick vector.

f. INFECTIONS KERATO-CONJUNCTIVITIS

This is a disease of the eyes of sheep caused by the organism Rickettsia conjunctiva in the presence of the bacterium known as Neisseria Ovia. It is found in Africa, Asia and Central and South America.

PHOTOSENSITIZATION

common in sheep in the topics where the ingestion of toxic plants may cause photosensitization and a high level of radiation aggravates the problem. Retention of affected sheep in dark sheds and proper feeding alleviates the problem.

FOOTROT

Although this disease is not usually fatal, it is of considerable economic importance and compounds the difficulties of managing sheep in wet regions, particularly in the humid topics. The disease could the caused by several organisms and could be handled by trimming the feet periodically, walking them through foot-baths containing 10% formaldehyde or 30% copper sulphate solution.

PARASITIC DISEASES

A. Trypanosomiasis: This is due to a protozoo called trypanosoma of which there are various species:

T. Vivax

T. Congolense

T. Brucei

These protozoa are carried by tsetse fly of the glosina species. The disease usually debilitates the animal by invading the central nervous

systems (CNS). However the West African dwarf sheep and goats are tolerant of the disease. Control is by:

a. Destruction of the insect:

b Spray of the trypancides e.g diminazine aceturate to

breeding and resting places.

c. Insecticides DDT for spraying

2. Chemotherapy- this method is onerous and must be repeated regularly.

3. Controlling growth of vegetation in areas which the animals are kept.

4. Biological control using sterile males.

B. GASTRO INTESTINAL HELMINTHIASIS.

These are caused by numerous species of parasites which are biologically distinct and exert defined pathogenic effects. The most important are the round warms referred to as:- i. Haemonchus contortus.

ii. Cestodes (Tape warm)

iii. Trematodes (Fluke warm) especially fasciola spp

iv. Schistomata

Helminthiasis causes retarded growth and loss of weight. Sometimes very difficult to differentiate their effects and those of malnutrition.

Control is by the use of anthelmintics

SELECTION OF BREEDING STOCK

Ewes that are too fat or too poor give a poor crop of

lambs. Ewes at mating time should be improving in condition and should have a little extra feeding but three weeks before the date of turning in the rams the flock should first be culled to remove sheep that are below standard. Feet should be attended to if any lameness is noticed. Ewes should always have access to mineral licks both on grazings and on folds.

Rams are smeared with a selected colour (specially bought for this purpose) between the forelegs so that they leave a mark on each ewe they serve. The coloring on the ram may change after about 2 weeks so that ewes not holding on the first mating will be contra-coloured and can be given another marking by the shepherd to remind him later of the delayed date of the lam bing.

Ewes that have not held to the first mating will return to ram after about 16 days. Rams are removed after 6 weeks and any ewes that remain uncoloured will be taken out and presumed to be non- breeders. They can be sold when fat. The selection of sheep for their ability to produce and rear lambs promises to be of immediate advantage and should be encouraged.

GESTATION PERIOD

____ The gestation period is 140-1 60 days (about 5 months) Preparation for actual lambing is usually made about 10 days before the end of the

J period counting from the date on which the ewes were serviced.

CASTRATION

____ This may be carried out using a knife or the elastrator as early as possible.

DOCKING

This is th3 cutting of the tail and is desirable in woolled breeds as this reduces the incidence of blowfly strike. Quresh (1968) investigated the influence of docking on a breed of fat-tailed sheep and found that the dressing percentage was higher in both males and females when they are docked.

WEANING

Weaning is usually accomplished at 4 to 5 months of age. Demiruren et al (1971) have shown that in Iranian breeds the rate of weight gain of lambs is significantly affected by the length of the suckling period up to

120 days from birth.

It is assumed that the teeth of early maturing breeds erupt at an earlier age than that of the late-maturing breeds, although there is no evidence to support this contention. The importance of this knowledge of noting

theemergence of the teeth as well as the degree of wear of the temporary to permanent stage is to look at the dentition and know the

ageof the animal.

ESTIMATION OF AGE BASED ON DENTITION:

Incisors Teeth

Two teeth (1st pair) 14-20 months

Four teeth (2 pair) 21-25 Months

Six teeth (4 pairs) 20-32 Months

I I Eight teeth (6 Pairs) over 32 Months

Sheep have four pairs of incisors each appearing at a different age. The I I table given above shows the respective age which each denotes.

GESTATION PERIOD

____ The gestation period is 140-1 60 days (about 5 months) Preparation for actual lambing is usually made about 10 days before the end of the

J period counting from the date on which the ewes were serviced.

CASTRATION

___ This may be carried out using a knife or the elastrator as early as possible.

DOCKING

This is th3 cutting of the tail and is desirable in woolled breeds as this reduces the incidence of blowfly strike. Quresh (1968) investigated the influence of docking on a breed of fat-tailed sheep and found that the dressing percentage was higher in both males and females when they are docked.

WEANING

Weaning is usually accomplished at 4 to 5 months of age. Demiruren et al (1971) have shown that in Iranian breeds the rate of weight gain of lambs is significantly affected by the length of the suckling period up to 120 days from birth.

It is assumed that the teeth of early maturing breeds erupt at an earlier age than that of the late-maturing breeds, although there is no evidence to support this contention. The importance of this knowledge of noting the emergence of the teeth as well as the degree of wear of the temporary to permanent stage is to look at the dentition and know the age of the animal.

ESTIMATION OF AGE BASED ON DENTITION:

Incisors Teeth

Two teeth (1st pair) 14-20 months

Four teeth (2 pair) 21-25 Months

Six teeth (4 pairs) 20-32 Months

I I Eight teeth (6 Pairs) over 32 Months

Sheep have four pairs of incisors each appearing at a different age. The I I table given above shows the respective age which each denotes.

c. RECORD KEEPING: Simple record keeping is preferred. A very popular system is by the uses of cards. See illustration of the card below:

Breed Remark

Birth Date

Twin or simple

Weaning weight

Age at weaning

Yearling weight

HENRY OJO

Chapter Three

GOATS PRODUCTION

The domestic goat belongs to the genus capra and includes five species:

C. hircus (The true goat)

C. ibex

C. caucasica

C. pyrenaica

C. falconeri

It is an important animal in subsistence agriculture on account of its unique ability to adapt and maintain itself in harsh environments.

Although the origin of domestic goats remains to be clearly established, the available evidence indicates that the Benzoar of South West Asia is the main ancestor.

Goats produce meat, milk, skin. In Nigeria the primary purpose is to produce meat and it is enjoyed in Nigeria. The population of sheep or goats in any part of the country depends on the preference of the people of such area.

Goat is a source of income to the farmers and they may be an integral part of a farm complex ,pecause of their inherent ability to utilize the products and by-productthe farm.

Goats are highly prolific-when they are adequately fed there are incidences of twinning and triplets. Profits are realized in a short time. In some places, goats have values for transport, prestige and sports. They are used as experimental animals in the study of digestion and physiology in the body.

Goat browse on all sorts of things e.g twigs, figs, and barks and herbs rich in crude fibber which sheep could not eat as a result fodder reserve which otherwise is hardly used by other animals are expected by goats. In may parts of the tropics there are definite limitations on the availability of nutrients for feeding dairy cattle. Under these conditions that are likely to be indefinite, the milking goat is a most useful animal.

The special attributes of goat's milk are that tubercie bacilli are rear, that there is a high proportion of smaller fat globules facilitating easy digestion and that it posseses anti-allergic properties. Its nutritive value does not differ appreciably from that of cow's mFlk (Parkash and Jenness 1968). Below is shown the average composition of goat's milk compared with the milk of Indian and European cows and Indian buffaloes.

In addition to their value as meat and milk, goat is also valued for the production of hair and skins. Skins are very important in India, Pakistan and several African countries. Another important product is mohair, the fleece of the Angora goat (plate 1.1) which is principally produced in South African, Lesotho and United States.

In addition to their main functions goats also serves as an investment against the failure of the cash crops, their ownership bestows prestige and in many communities they have a place in local custom, religion and festive . :&rsions. They are also used for the production of manure.

There has been some disagreement as to the value of goats because of the widely held belief that they do damage to trees and vegetation especially in arid regions. The fact remains that this bad reputation stems more from their mismanagement than from any inherent fault. With controlled management they can be a great help in agricultural development and food production.

DISTRIBUTION

The total population of goats is approximately 508 300,000. About 54 per cent of this is found in the tropics. The largest concentrations are in African and the Indian sub-continent.

Goats are possibly the most widely distributed of domestic livestock. They are found in countries representing the climatic extremes of South America to the wet and humid tropics of south Asia. Their rustic and hardy qualities enable them to withstand dry environmental conditions much better than cattle. They perform best however, in the drier tropics and on light sandy soil. In Africa the greatest concentration of goats are to be: found in East Africa, Northern Nigeria and Morocco. This pattern is also applicable in Indian subcontinent, Western Asia, South and Central America and the Caribbean.

NUBIAN

Associated with Southern Sudan but is widespread in North-East Africa and the Mediterranean coastal belt. It is unique in being the only African breed specialized for milk production. It is the main progenitor of the Anglo-Nubian breed. Possesses large udder.

Nubian is a large long-legged goat with long pendulous ears and a pronounced Roman nose at least in the male. Horns are present in both sexes in some strains but absent in others. Some strains are predomin3ntly black, and others brown it has long hairs. Withers height is around 70 — 80 cm, and live weight is around 27 — 60 kg. Milk yields are in the order of 1-2 kg daily or 120 to 140 kg annually in two lactations.

RED SOKOTO

Uniformly dark red in colour, horned in both sexes, and short haired.

The ears are short and carried horizontally. It is very local in distribution being confined to the Republic of Niger to Sokoto State in Nigeria.

It is relatively small (23-30kg). The skin of Red Sokoto is among the most valuable of all goat skins and the breed is also a good meat animal. Gives daily milk yields around 0.5 kg in dry season, and up to 1.5 in the wet season.

WEST AFRICAN DWARF GOATS

Goats of this type, with disproportionately short legs which are often bent, occur in and near the tropical forest belt in West and Central African. These are about 50cm in height and 20kg in weight tending to be larger as the Savannah Zone is approached.

Growth rate and milk yield are very low, but twin and triplet births are common and they breed at all times of the year. They are used almost exclusively for meant production, and the skin which has little commercial value is usually eaten.

USEFUL PECULIARITIES

Adapted to the humid tropical environment to resistance to trypanosomiasis

BOER GOAT

South African breed, intensive selection has been made both for colour pattern and meat production. Improved Boers are white with red head markings, and when well fed are of excellent meat conformation, their skins are also valuable, fertility is high and milk yield fairly good. They appear to have been developed from local African goats of the long- legged type, crossed with goats imported from India to the Near East. They bear some resemblance to Nubian goats having lop ears and markedly convex noses. They are very hardy under tropical or subtropical range grazing conditions but are not suited to humid regions. Twin or triplet births are usual, and under good husbandry conditions milk yield is sufficient to enable kids to reach live weights of around 40kg by 12 months of age or less.

ANGORA GOAT

Originated in central Asia. Introduced to South Africa in 1838 to the United States in 1849.

Angora goats have also been introduced, mainly for cross breeding with indigenous long haired goats into the Soviet Union, India, Pakistan, Fiji, Madagascar, and Australia.

Angoras thrive best in hot, dry sub-tropical climates such as that in central Turkey, South Africa, and Texas. Although they are bred mainly for the mohair, meat and milk are subsidiary products'. The hair grows in long white lustrous locks or vinglets usually from 13-15cm long and up to 25cm. in U.S.A. where the goats may be shorn twice, the average annual yield is 2.9kg.

Angora goats have been successfully used for cross breeding with various indigenous tropical breeds to develop mohair production. This has been done in South Africa and Lesotho, and also in India, Madagascar and Fiji.

DAMASCUS

Most suitable dairy goat in the middle East. The breed is common in Lebanon, Syria and Cyprus.

It is polled in both sexes generally red or red and white in colour with long lop ears. Heights at withers 70-75cm, and live weight 40-60kg. Daily milk yield 2-4 litres with total lactation yields of 300-600 litres in about eight months.

JAM NAPARI

India to Pakistan. Most popular and widespread breed in use for milk production in India, South East Asia.

Jamnaparis are excellent milk animals and are also often used for meat production. They may be of various colours including white, tan and black. The ears are pendulous and about 30cm long. The udders are

Well developed. Males weight about 68-9 1 kg and females is 96-63kg live weight. The height at withers of males and females is 91-127 to 76-92cm

respectively. The approximate milk production is 235kg over a lactation period of 261 days. In India as much as 3.8kg per day and a maximum yield of about 562kg have been recorded. In view of its excellent milking abilities and its growth potential, the breed has been widely used to upgrade smaller indigenous goats in countries such as the West Indies, Malaysia and Indonesia.

EUROPEAN BREEDS IN THE TROPICS

SAANEN

Originated in West Switzerland. They are white pale cream or pale biscuit in colour, with black spots on the nose, ears and the udder. They have a short coat and are generally polled. Ears are erect and point forward. They are good milkers and have been introduced into Puerto Rico, the West Indies, Fiji, Ghana, Kenya, Malayisa, and Australia. They are the main breed of milking goats in Australia. They tend to be

sensitive to strong sunlight and there is the important need, therefore to shade them from the sun and give them good, indoor management.

Saanen goats have been very popular in the West Indies and lactation yields of about 800kg over 250 days have been achieved.

BREEDING

Sexual maturity in the goat is achieved at 4-6 mouths of age, but management practices are offen designed to delay mating until the does

are near to mature body weight, so that pregnancy does not coincide with the period when the does are actively growing. The tendency is therefore to mate does at about 12 moths so that they kid for the first time at about 18 mouths of age.

The duration of the estrous cycle is 18-21 day, same as in sheep and the duration of estrus is about 24-36 hours. There could be variation. Gestation period (duration of pregnancy) has been found to bee fairly constant at about 146 days, with a range of from 145-148 days.

MANAGEMENT OF BREEDING STOCK

Under any system of husbandry it is necessary to be able to divide the herd into more than one group and to keep them securely apart. If breeding is not to be restricted to any particular season and male production for human consumption is not required, the bucks may run with the adult female at all times. It is then only essential to separate the young females from 3 mouths old until they are considered old enough for breeding. If breeding is to be restricted to certain seasons or several bucks used each with its own group of females, either these subherds must be kept separately housed or the bucks kept housed while the does all graze together by day but are divided into mating groups each night and housed with appropriate buck.

KID REARING ft eb-'l t frik tIL Qp) since it stimulates the alimentary canal and also contain vitamin A and

antibodies which confer Immunity against diseases. If

the kids are not sucking within hours of birth it may be necessary to teach them to do so by diverting milk from the teats to the mouth after examining the doe's udder and teats to make sure there is no abnormality. The kids begin to nibble at solid food such as bush leaves, gasses or dry fodder when they are about 2-3 weeks old and these should be encouraged.

CASTRATION OF MALE KIDS

Matekids not required for breeding should be castrated. This improves the quality of the meat. Castration should be done as soon after birth as

practicable. It can be done in one or 2 ways- bloodless castration or with a knife.

• HOUSING

No elaborate housing is necessary, but what is provided should be light well ventilated, well drained and easily cleaned. Two types of housing

are commonly used:-

Leanto -type or the ground — level house common in most parts of the tropics. This measures about 2 to 3m high in front and sloping to 1 .5m at the back or with the sloping eaves standing out by about 0.5m. The floor can be rough concrete or more commonly rammed clay or earth. The other type is the stilted housing where floor is raised I to I .5m above the ground level for easy cleaning and collection of dung and urine.

FLOOR SPACING

1.8 x 1 .8m could take or house 10 kids, 2.4 x 1 .8m for 10 bucks. If you

are housing 100 — 125 herds of goats the floor spacing should be 12 x

18m.

MANAGEMENT

In any system of management, the main idea is to control the

environment and the breeding stock..

(a) Control of environment: Provide adequate housing facilities to combat environment hazards.

(b) Improve the breeding stock: Know the breeding system to adopt. Separate your breeding stock if you are not practicing indiscriminate breeding.

SYSTEM OF MANAGEMENT

1. INTENSIVE SYSTEM: This system is mainly for Zero grazing. It is the system for those who would be productive in their husbandry. It gives complete control over the destructive aspects of the goats' feeding habits. Additionally feed the animals with

concentrates.

2. SEMI- INTENSIVE SYSTEM: This system involves controlled grazing of fenced pasture with some supplementary

concentrates feeding.

3. EXTENSIVE SYSTEM: In the extensive system housing is

providedat night and during rains. Also provide grazing grounds. This system is not good for dairy goats. A land not

suitablefor agriculture is good for this system.

GENETIC IMPROVEMENT

Over wide areas of the tropics there is little or no attempt to restrict reproduction to select any individual, the animal being permitted to breed indiscriminately apart from perhaps the occasional introduction of a large or attractively marked male.

Rapid genetic improvement can be obtained by supplying farmers with improved bucks from recorded parents. The commercial breeder however needs to cull the defective and aged females and the productivity of his herds will increase provided h.usbandry is adequate and a succession of improved bucks is available from the breeding centers.

The temptation to use his own males sired by improved bucks must be resisted and all males in commercial herds should be castrated.

Introduction of systematic culling will be advantageous both to commercial herd and studd breeders aimed at eliminating young females of excessively small size, removing females that fail to rear the required minimum number of kids annually and those of low milk yield in dairy herds and culling all females in which productivity is declining due to age.

FERTILITY AND PROLIFICACY

An animal is fertile if it produces the normal spermatozoa or ova capable of fertilization. An animal is said to be prolific if it produces numerous offspring. Since a prolific animal must also be highly fertile, it is not surprising that the terms are used synonymously. Prolificacy of the female goats may be expressed as the number of kids born per doe (or per 100 does per year) or as number of kids per birth (litter size).

While prolificacy is a useful indication of the material ability of the doe, the number of kids reared to weaning (reproductive efficiency) is of greater practical importance since a doe which although prolific fails to rear a kid is worthless.

NUTRITION AND FEEDING

Goats have typical habits which are typical of their species as a whole. Goats are able to graze on very short grasses and browse on foliage which are not normally eaten by other domestic livestock. They are inquisitive feeders with a feed range from herbage and tree bark to skins and cloth. More than half the browsing time is occupied eating leaves and shoots of trees and bushes.

NUTRITIONAL REQUIREMENT

The energy requirements for maintenance in goats are similar to those of sheep, being 725.8g starch equipment (SE) per day per 100kg live

weight. The energy requirement for live weight gain is 3.0g SE per kg live weight gain. The digestible crude protein (DCP) requirement for

maintenance and for milk production are 45 to 64g per 100kg live weight and .70g per litre of milk respectively.

The mineral and vitamin requirements of goats are equally important. Goats in milk production have a very high reql.lirement for sodium chloride, and salt licks and I or mixed mineral licks should be made available. Adequate water supplies is very important.

The nutritional needs of goats should be done realistically and should be based on cheap foods such as browse, pasture and agricultural and industrial by-products and crop residues such as rice straw, sweet potato vines and cassava should be fully exploited. Full advantage must be taken of the ability of goats to digest cellulose particularly well, as they cannot compete with pigs and poultry in efficiency of conversion of concentrates to protein foods.

Whatever should be fed to the goats as concentrates should depend on what is available, and in the tropics the variety is very great.

DIFFERENCES BETWEEN SHEEP AND GOATS

1. Goats are horned in both sexes although the Anglo Nubian female goats is polled.

2. Arrangement of teeth in the goats is different from sheep which could account for the grazing habit of the goat.

3. Sheep is more heavily manned than the goats

4. Tail of goats is curled upwards while that of sheep is downward.

5. Hair of sheep is longer than that of goats with the exception of the Angoran goats.

6. Presence of Iachrymal glands in the skull of the goats absent in the sheep.

7. The ram shows more aggression in defense which the buck is lacking.

8. The male goats has odour.

POTENTIAL FOR INCREASED RODUCTION

With proper management the potential for increased production from goats is considerable. Much will depend on whether their value as small domestic animals is adequately recognized. Small size is significant on account of their economic, managerial, and biological advantages. Attendant to these features is their high fertility and short generation interval, which means that milk production begins 5 moiths after initial mating and the first c'rcase may be on sale in less than a year.

Two aspects that merit special mention and may be worthy of exploitation are the higher efficiency for cellulose digestion and food utilization for milk production of goats. Both are important, as they raise

questions as to the role of goats in increasing meat and milk production. Evidence from Pakistan (Wahid,

1965) and from. Greece and Cyprus (French, 1970) suggests that goats are relatively more efficient and economic than some other ruminants in this role. The prospects for increased productivity, based on exploiting these special features inherent in the goat, should have the objectives of increasing the number and size of kids and frequency of kidding and lengthening the productive life span.

Chapter Four

PIG PRODUCTION

The primary purpose of pig production all over the world is for meat, including pork, bacon or fat. The secondary considerations are the production of pig skin, bristles and manure.

ADVANTAGES OF PIG FARMING

1. Pig farming can be carried out in almost any type of land e.g. good or bad land, hilly, arid or even rough land.

2. Only a small amount of land is required.

3. Pig production is not subject to weather variation as in agriculture and horticulture.

4. Much more amenable to accurate planning, budgeting, and forecasting than most other types of farming.

5. Large scale specialized swine production is almost

like a factory procedure of the conveyor-belt type.

6. Because of high fecundity and rapid growth rate, rapid rate of return is assured.

USES OF PIG

1. As a source of meat in U.S.A. and Africa and can be cured in various ways as bacon, smoked, salted or sausages.

2. Surplus fat made into lard (cooking oil).

3. Manufacture of industrial goods-bookbinding leather, hard bags, brushes briefcase, etc.

4. Manure production.

5. Pigs convert waste products into good quality pork.

6. Provision of a market for feed.

STARTING POINT, MANAGEMENT AND HOUSING

The successful raising or husbandry of pig depends on sound planning based on the known biological or genetic make up of the pig. The aim of modern pig farming is to produce meat at as low a cost as possible.

The pig as a non-sweating species is very sensitive to changes in climatic environment. Any management system envisaged must therefore take these factors into consideration in order to secure maximum productivity. The fact that modern pig and breeds are derived at least partially from a wild species that was adapted to a warm, shaded humid environment tends to give the impression that pigs in the tropics can be managed in or outdoors so long as adequate shade, wallowing system, drinking water and large quantities of feed stuff are available.

Unfortunately one major constraint in managing pigs outdoors is the very incidence of certain parasites as the humid tropics is their ideal habitat (kidney wcrm-stephanurus dentatus). Rotation of the pigs and vermifuge drugs are used to eradicate the worms.

Two types of management systems are known:

a. The small scale or peasant producer.

b. Large-scale producer or commercial production.

We shall make do with the small-scale production for now.

a. The peasant producer (small scale producer): In many tropical villages pigs are left running around as scavengers. They pick up offals where they can and if ever any conscious effort is made at feeding them, waste foods such as rice bran and root peelings are provided. Little attention is given to the breeding stock with the result that some times the runt in the litter is selected to raise the future sire.

In order to improve productivity under this system of management the following steps are to be recommended:

j. Where land is available and in abundance pigs could be fenced in small paddocks in which roots such as cassava, sweet potatoes and yams are also grown and cultivated for their use. It would be essential to provide shades in the paddock. There should be adequate water supply and if possible the paddocks should be surrounded by wire-netting. Alternatively bamboos or coconut trunks or live poles can be used for fencing even though they are less permanent structures. The pigs raised under this condition may not grow rapidly but at least evidence abound that they grow a lot faster than those left roaming loose in the village.

1. Pens could be constructed of rough timber with a thatched roof and concrete floor. If built close to stream, water could be diverted to run through them. The running water could then serve as drinking water for the pigs also used for cleaning up of pens and wallowing. Feeds such as offal and roots would then be made available in the pens. Other unconventional feeds are rice bran, palm kernel cake, cotton seed cake, gleaning from cowpea farm, bark of banana etó.

j. Pens could be built of rough timber, block wall, thatched roof or asbestos, concrete floor or earth floor and coarse hay straw continuously thrown in.

FARROWING MANAGEMENT

1. For a beginner it is better to start with in-sows.

2. Whatever the system the gilt or sow should be farrowed in a separate pen (separated one month before farrowing)

3. Straw to be provided for her one week before farrowing.

4. Give extra feed towards end of pregnancy and reduce one week to farrowin g.

5. Use farrowing crates.

6. Gestations period is 114-116 days (3 months, 3 weeks and 3 days)

7. Sows shed up 20 eggs during ovulation but mature sows farrow 8-

14 piglets per litter. Mate twice if possible with a different boar.

8. It is possible for a sow to produce two or more litters a year.

FARROWING HAZARD

1. Mastitis and agalactia: If this happens foster the piglets but within limits of three days to ensure administration of cholestrum.

2. Remove afterbirth to avoid cannibalism.

3. There could be still-birth, over laying, cannibalism, pneumonia, enteritis, anemia, scours.

4. Apply good management to neutralize these factors.

MANAGEMENT OF BREEDING STOCK

Guts should be selected for the breeding herd at 4-5 months of age, when up to 68-91kg. Where it is possible they should be selected on basis of records to ensure that they do not possess any inherited defects and that they come from families noted for large litters and sexual maturity. They should be healthy, possess sound feet, well grown with at least fourteen prominent teats and a good carcase conformation. It should be one or two boars to every 50 gilts so more care should be taken in selecting boar.

Guts can he served at the age of 7-8 months. The SOW'S greatest fertility occurs at mid-oestrus (about the time of ovulation). For greater degree of certainty of fertilization the sow should be mated twice during estrus i.e. 12 and 36 hours after its onset. The interval between estrus period is approximately 21 days and gilts tend to have shorter periods than the sows. Boars mature at 8 months for breeding. Breeding seemshbe at all season.

CASTRATION

All males (boars) destined for market should be castrated 6-8 weeks.

REPRODUCTIVE BEHAVIOUR

Gilts tend to have a shorter heat period than sows. Within this cycle heat period lasts up to 48 hours. Females in heat are characterized by grunting, restlessness and swelling and redness of vulva. Gilts should be bred for the first time on or after their third heat period while for sows it should be first or second heat after weaning.

PIGLET MORTALITY

Research has shown that piglet mortality in the tropics is of the order as in the temperate zone (Davids, 1948). The major mortality was caused by sow overlaying. Suitable farrowing equipment is the answer. This overlaying is less in the first and second litters. Farmers view of culling after this period is irrational. Ideal thing will be to invest in properly designed farrowing houses.

PIGLET ANAEMIA

Pigs are born with relatively small reserve of iron for their body and their mother's milk does not normally provide sufficient iron for their requirements. Signs are paleness in the region of the ears and belly, listlessness, rapid breath and diarrhea. As a remedy place clean fresh earth in the piglets' pen each day or give iron injection (100mg Iron Dextran) 3 days after birth.

SUCKLING AND CREEP FEEDING

Because of the problem associated with the number of piglets to the number of teats in the sow's udder, the piglets that could not be accommodated are fostered or raised by hand. At one week of age creep feeds should be introduced -a practice of feeding piglets separate from their mother. Feeding is 0.7 — 1.4kg per day per pig.

WEANING AND WEANERS

At weaning the sow should be taken away from the piglets and not the reverse. It should be gradual. Weaning should be done at 8 weeks. The average number weaned and the weight at weaning is mainly determined by the standard of management and feeding and not breed characteristic. Feeding is 0.7-1.4kg per day per pig.

FATTENERS

Pigs raised in the tropics are raised for pork and generally slaughtered at

45-63kg liyeweight. Pigs are usually fattened in groups of the same age.

An avera&J2.7kg/day/pig is adequate.

HOUSING

All pigs in this our context should be raised on concrete floors. The floor should not be too smooth to avoid skidding. Concrete floors adequately controls internal parasite and reduces labor cost. A series of pen simple and flexible that can be adapted for farrowing, fattening, or breeding stock are needed. Some differences in pen size if desired increases the flexibility of the system. Pens could be equipped with wallows, farrowing crate, feeding troughs, watering compartment, creep feeding facilitates and drainage pipes. A pen of 2.4x4m will conveniently accommodate a sow and her litter, up to twelve porker pigs, eight bacon pigs or three breeding sows.

NUTRITION

Nutrition of swine in the tropics not considered vastly different from the nutrition of pigs in temperate climate. The difference lies in the effect of our weather on the feeds in form of moulds, fungi and product deterioration.

Proper and adequate feeding in swine production are considered important since feed represents 80% of total cost of production. This is because pigs grow rapidly and therefore its food demand becomes very high for instance a baby pig may weigh 1.4kg at birth and about 160kg or more just in 18 months later. Unfortunately pigs thrive best on just those foods which are available to humans but fortunately can also thrive on by-products and other materials not useful to man.

The general aim therefore is to use these cheaper, lower grade feeding stuff to the fullest extent supplemented by the more expensive, more nutritious feed to the point where true and sensible economy exists.

The major problem in swine feeding in the tropics therefore relates to the efficient use of unconventional feed tuff such, as cassava, yam, bananas, plantains, rice, by-products, sugar cane by-products and protein supplemental sources, potatoes, oil cakes, carcase and fish byproducts, milk and milk by-products, pasture grass, yeast, antibiotics, banana barks and wastes from catering industry.

DISEASES

One major problem confronting pig producers in the tropics is high mortality resulting from diseases. In

Philippines mortality is as high as 50% and 30% in Fiji.

Some diseases such as Brucellosis, leptospirosis, mastitis, Agalactia are found in breeding stock. Others are transmis ble gastroenteritis (TGE), swine influenza, Africa Swine Fever (ASF) or Hog Cholera, Swine plague, mite, Lice, internal parasites (lung worms, roundworms, nodular worms, earthworms, whipworms, kidney worms).

CONTROL

The most effective control measure against diseases is preventive action. Preventive action will reduce nutritional, climatic and other environmental stresses to a minimum by good management.

NOTE THE FOLLOWING

1. Use vaccination for cases requiring vaccination.

2. Control internal and external parasites by spraying/drenching when necessary.

3. Adequate feeding at all stages and a good wallow (Wateer Spray hardly possible here)

4. Segregate animals diseased in an outbreak.

5. Dispose dead pigs properly.

6. Cleaning and disinfecting of all premises and

equipment after outbreak and non-use for 4 weeks.

BREEDS OF PIGS

Table

BRITISH WEST AFRICAN BREED

Birkshire

Large Black

Large White

Middle Sex

Tamworth

Bakosi (Camëroun)

OTHER BREEDS

Land race

C rao n

American Breeds

PHILIPPINES

Duroc Jerseys

Ilocos

Hammpshire

Jalajaja

Poland China

koronadal

CONSEQUENCES OF BAD MANAGEMENT

1. You could loose your stock and go out of business

2. You should love your animals and follow all the approved practices as advised

3. It is an area you know little or nothing about so depend on the advice of experts.

4. It can cause hypertension

5. Loss of Credibility

6. You rob future generations of the skill.

7. Sponsoring agency may be discouraged. A bad report could be written about the project so effort should be made to avoid failure.

8. Nigerian government removed her hands from direct production because of failure of past projects. You cannot afford to fail.

RECENT ADVANCES IN SWINE PRODUCTION

1. Genetic engineering

2. Artificial Insemination

3. Estrus Synchronization

4. Super-ovulation

5. Conveyor-belt type production.

CONCLUSION

Past efforts by both government and agricultural agencies had always geared towards crop production. This had been at the detriment of livestock production. The low level Of animals based protein consumption in the country had wrecked havoc on the. mental and physical development of Nigerians. However, it is not too late to make amends; to correct the ills of the past.

Chapter Five

RABBIT PRODUCTION IN THE TROPICS

Rabbit is one of the small animals which look attractive and easy to handle. This animal has vast advantages over other animals. Unfortunately many people have not known these vast advantage and have not benefited from their immense benefits.

However, whoever keeps rabbits should ensure that the rabbits are not accessible to dogs or soldier ants (adequate housing). They should also ensure that they acquire the required skill, feed the rabbits adequately and also work hard to ensure there are no lapses.

A beginner can start with just a female rabbit (a doe), and a male rabbit (buck) which is cheap enough for anybody to invest in.

REASONS FOR KEEPING RABBITS

The two main reason for keeping rabbits in the tropics

are for food purposes and also for monetary reason. However, there are other scientific and social reasons for keeping rabbits:

(a) Rabbits require small investment for beginners.

(b) Rabbits keeping is a low cost, high quality protein that uses forages and food wastes of no value to human.

(C) The venture is not restricted by taboos and religious beliefs.

(d) In emergencies rabbits can be raised in a short interval and its meat (veal) has a high lean quality.

(e) Ideal for school farm.

(f) It is a good animal with which to learn the basic skills of animal husbandry since it can withstand the sometime rough hadljof beginners.

(g) It is not smelly or noisy animal and can easily be kept near school buildings and homes of individual.

(h) It is an attractive animal for handicapped children who enjoy in

its care and management

HANDLING OF RABBITS

Correct rabbit handling is very important because rough handling is a

cause of stress to the rabbit, and may also be cruel. The skeleton of the

rabbit is not as heavy or as strong as for example, that of cat. Its

backbone is easily damaged-often leading to paralysis-if it is dropped; hence the need for skillful handling. A board ora piece of card can be used to trap the rabbit in a corner, to avoid chasing it around. Simply holding the base of the ears whilst covering the face and eyes will usually calm the rabbit. Do not lift rabbits by their ears because it is very painful for them and can cause the stretching of the ear at their bases which make them to droop when they should be erect.

Rabbits are good listeners, so you could also talk to them in a soothing voice when handling them.

Basically you can handle rabbits

(a) By the scruff — skin behind the ears

(b) By the pelvis — usually for young rabbits

ENVIRONMENTAL FACTORS AFFECTING RABBIT PRODUCTION

The Rabbits Environment

Rabbits interacts with its environment. The major environmental factors are:

(a) Hutch microclimate

(b) Floor and hutch space available for rabbit

(c) The presence and number of other rabbits

(d) Parasites and disease organisms

(e) The rabbit keeper

(f) Other things near the hutch e.g. mouse, dogs, rats etc.

Homeothermy

The rabbit is a homeotherm. This means it must keep its body temperature within certain limits if it. is to remain alive. The rabbits

normal body temperature (Rectal) is 37-39.5°C (99 — 103°F). It maintains this temperature by burning or breaking down food in its body and if necessary, by working to keep wc.rm e.g. by shive

HEAT STRESS

Ambient heat constitutes a real problems for rabbits and is a significant constraint to their successful production. The ears of rabbits are well supplied with blood and act as radiators and convectors of heat. Apart from that the rabbit is not adapted for rapid heat loss.

Heat stress could arise due to sudden rise in temperature, exposure to direct rays of the sun, where hutches are made partly or wholly of tin, where there is no shade or air movement or where humidity is high. Heat stress is a problem in the production of rabbits in the tropics. Until truly tropical breeds are developed it remains an essentially temperate animal which is better able to cope with cold then heat.

NUTRITIONAL PHYSIOLOGY

INTRODUCTION

The nutrition of the rabbit is the single most important aspect of production. Well — fed rabbits can better resist diseases and recover from environmental stresses and mistakes made by the rabbit keeper.

THE DIGESTIVE SYSTEM

The digestive system can be described under three headings:

(a) Mouth and teeth

(b) Stomach and small intestine

I I (C) Caecum and large intestine

Mouth and Teeth:- This segment of the digestive system consists of the I I incisors, the canines, the premolars and molars. The incisors are for

cutting and premolar and molar for grinding.

I I The food is mixed with saliva as it is ground by the molars to reduce its

particle size. After this first processing period the food is swallowed and

passes down the sophagus to the stomach.

STOMACH AND INTESTINE

The stomach represent approximately 40% of the total volume of the digestive system. Food in the stomach is exposed to acidity and some enzymes digestion begins. Weak muscular contractions in the stomach push the food into the first loop of the small intestine, the duodenum.

The food is first bathed in bile which enters via the bile duct. Bile is produced in the liver and is stored in the gull bladder. The bile salts assist in the digestion of fats in the food.

As the food passes further along the duodenum it is mixed with enzymes produced in the pancreas and which enter via the pancreatic duct.

Enzymes digestion is rapid and food proteins are broken down to amino acid which are absorbed through the intestinal wall into the blood system. Fatty acids, glycerol, glucose and other simple sugars are the end products of fat and carbohydrate digestion and these are also absorbed as the food passes along the small intestine.

CAECUM AND LARGE INTESTINE

I I The rabbit is sometimes referred to as a hind - gut fermenter, meaning that food is broken down by bacteria at the end of the digestive system. The major site of this breakdown is the caecum. The large caecum has I I absorbing and secretory cells throughout its large area. At its end there is a small closed sac, the appendix.

The caecum contains many bacteria that grow and multiply on the partly — digested food. These bacteria are very important because they I I synthesis B vitamins, particularly thiamin, and because they break down plant fibre. This breakdown results in the production of acetic, propionic and butyric fatty acids, which are absorbed from the caecum and large intestine and used as sources of energy by the rabbit.

I I Water is eabsorbed throughout the caecum and large intestine. This results in the relatively hard, dry pellets which are characteristic of rabbit faeces.

NUTRITIONAL REQUIREMENTS

Water may not technically be a nutrient but it is nonetheless an essential requirement. Water consumption in rabbits is greater than might be anticipated. This is especially so for the lactating doe

Like all animals, rabbits need four major groups of nutrients:

(a) Carbohydrates and fats

(b) Proteins

(C) Minerals

(d) Vitamins

CARBOHYDRATES AND FATS:- Carbohydrates and fats provide energy. Energy is the rabbit's petrol. Without petrol a car cannot run. Similarly, without energy a rabbit will soon die. Rabbits derive their energy from plants and plant products such as carbohydrates, starch, cellulose, fats and other substances.

Energy is used to contract muscles whichenable the rabbit move. It is also used to join substances together to build up the rabbit body and to make products such as hair and milk. Energy-concentrated food and not forage alone is what the rabbit needs. Forages are dilute sources of energy.

PROTEIN:- All the body tissues other than bone, teeth and fat (eg muscle, hair, skin) aLe proteins. Enzymes and hormones - important body chemicals arecalled amino acids. The rabbit makes its own

40

particular protein from proteins it obtains from its food. There are over 200 amino acids known in nature, but there are only about 20 which occur widely in plants and animal tissues. Ten of these are regarded as essential for the rabbit. They are:

- Lysine

- Methionine

- Arginine

- Phenylalanine

- Histidine
- Valine
- Threonine
- Tryptophan
- Leucine
- Iso leucine

The essential amino acids cannot be synthesized by the rabbits as can the other non-essential amino acids. The two most important in the above list are lysine and methionine as these are usually amino acids which are most likely to be deficient in rabbit diet.

For rabbits the recommended crude protein level in the dry matter of the ration is

- Over 18% for newly weaned rabbits;
- 16-18% for rabbits from 12 - 24 weeks;
- 15-17% for breeding does

MINERALS

Most of the minerals in the rabbits body are in the bones and teeth, which contain large amount of the two mineral, calcium and phosphorous. The mineral that rabbits require are divided into two categories:

major and trace. Major minerals are required in relatively large amcunts and the minor minerals are all other minerals which are required in relatively small amounts as listed in the Table 1.

Calcium and phosphorous help to give the bones their hardness. They are also involved in maintaining the acid alkaline balance in the blood. Phosphorus is also involved in energy transfer within the body cells. Calcium, phosphorous and vitamin D are often considered together because they interit with each other.

Table Major and Trace Minerals Important in Rabbit Nutrition

Major Minerals

Trace Minerals

Calcium

Iron

Phosphorous

Copper

Magnesium

Sulphur

Sodium

Cobalt

Potassium

Zinc

Chlorine

Manganese

Selenium

Iodine

VITAMINS:- Vitamin are chemical that are required in very small amounts to speed up chemical reactions within the rabbits body. The most important vitamin are A and D, and the B vitamins chlorine and thiamin.

BREEDS

Choosing a Breed:- In selecting a breed many factors have to be considered.

Availability, personal likeness, suitability for local market condition, cost and other factors of economic consideration affect success in commercial rabbit industry.

Availability of breeding stock is the most important consideration in breed selection. If the breed chosen is in plentiful supply in immediate locality foundation stock will be available from many breeders.

Selecting a breed or strain should be done on the basis of body conformation, pedigree and performance records. Progeny records where available are definitely most satisfactory indications of the family's value but where these are not available the pedigree and performance records of the individual's close relatives should be studied.

Attention to body conformation particularly of young fattening rabbits, though not as valuable as progeny record, offers the prospective purchaser an indication of the type of rabbit that can be bred from such stock.

COMMON BREEDS OF RABBITS

NEW ZEALAND WHITE

This breed is the one used most widely throughout the world for meat production. It is all white in colour and usually weighs. 3 — 5kg when mature.

CALl FORN IAN

This is the second most popular breed for meat production. The colour is all white but with tipping on the nose, ears and tail. The weight range for the mature Californian is 3 — 4.5kg.

DUTCH

The Dutch is a small compact meat rabbit of 2.5 — 3.5kg. It has a wide band of fur around its body at the shoUlders as well as a white stripe down the middle of its face. Its front feet fall within the white band. The tips of its back feet are also white. It has a high meat to bone ratio and mature between 4 /2 - 5 1/2 mo ths It has a low maintenance requirement compared to other rabbits.

CHINCHILLA GIGANTA

This is a large breed which matures at a comparatively early age. The breed is blue — grey in colour with a white belly. It has a characteristic ruff or dewlap. This is a thick fold of skin around the front of the chest which is very obvious when the rabbit is in good condition. The weight range for the breed when mature is 3 — 4.5kg.

NEW ZEALAND RED

This breed is essentially a red New Zealand white type which has not been intensively selected for growth rate. Mature live weight of 3 — 4.5kg is lower than the white type.

FLEMISH GIANT

As its name goes this is a giant breed which at maturity

can weigh well over 6kg. It is usually light grey in colour but may also be sandy, blue or white. This is not a suitable breed for the beginner.

OTHER BREEDS

There are over 40 breeds of domesticated rabbit in the world. Other common breeds include Angora, Champaign, D' Argent and Burgundy Fawn.

BREEDIN:.G

SEXING AND EARLY GROWTH

To determine the sex of rabbits is not difficult, with a little practice. This can be carried out shortly after weaning at 6 — 8 weeks. This is the time when the males and females should be separated and kept in separate hutches. For sexing, the rabbit should be held on a flat surface. The skin around the genital opening should be gently pushed back with the finger and thumb. In the male this will reveal the penis as a rounded tube-like structure. In the female the vulva will be exposed as an oval opening to I the reproductive tract.

If any difficulty arises in differentiating the sexes examine the genitalia of

I fully grown rabbits.

During the early period of growth, care must be taken to provide as good quality food as possible. Low quality forages should be avoided especially for growing rabbits otherwise it would result to stunted growth.

I SELECTION FOR BREEDING

At the age of 4 — 5 months, rabbits should be selected for future i breeding stock. The following criteria should be used:I (a) The live weight of the rabbits: Select the heaviest, taking into consideration differences in age when you select from more I than one litter.

(b) The size of the litter from which the rabbit came from.

(c) Healthiness — select very healthy rabbits.

(d) Males should be selected on the basis of the above making sure they have two testicles in their scrotum.

I MATING

When feeding mainly on forage, does are normally ready for mating at I about 8 — 10 months and bucks 6 — 8 months depending however on condition. The ratio of does should be in the range of 10:1. Itis however not out of place to have two using one as a standby.

It is advisable to mate early in the morning or evening. Avoid mating during the hot periods of the day. Take the doe to the buck and never Ithe reverse +- avoid attack from the doe.

The doe stand still within few seconds if she is ready for mating,

Isecondly raises her hindquarters in order to allow the buck to mount and mate. Successful mating is signaled by the buck thrusting forward and literally falling off the doe. As soon as mating has occurred the doe I should be removed from the bucks hutch. The.doe may be returned after an hour or so for a second mating.

I If the doe does not stand for the buck or if she attacks him it is useless to persist. Return the doe to her apartment and make a second attempt.

I Always allow the buck a period of rest after a successful mating before introducing another doe.

Indication of Estrus: In estrus the vulva is sometimes swollen, chin rubbing and restlessness are sometimes observed.

PSEUDO PREGNANCY

Following a mating between an apparently healthy doe and buck ova are released from the ovary 8 — 13 hours after. In many cases however conception fails to occur, the ova merely being liberated but not fertilized. This phenomenon is controlled by hormones in the body.

There is the occurrence of hormonal changes usually associated with normal pregnancy. This is false pregnancy. A doe which has more than one false pregnancy should be clled.

GESTATION PERIOD (PERIOD OF PREGNANCY)

The duration of pregnancy varies with strains but is usually between 30 — 31 days after fertility. Most does kindle on the 31st day after fertilization.

DETERMINING PREGNANCY

In the commercial rabbitary it is extremely important to know whether maternal does are pregnant or otherwise. Not only can time and money be saved by knowing which does are pregnant but also the correct feed plan for does in either state. In this regard a correct feeding regime can be carried out for the benefit of both pregnant and non pregnant does.

(a). PERIODIC WEIGHING OF THE DOE

Any gain in weight is taken to indicate the extent of developing fetus. This method is only moderately satisfactory when carried out during the final week of pregnancy period. Even at this late stage a gain weight is not conclusively evidence of complete pregnancy due to varying response of many does to different feeding programmes and management techniques.

(b). PALPATION TECHNIQUES

This is the most reliable method. It involves feeling with the thumb and fingers the developing embryo in the horn of the uterus.

The does should preferably be removed from her normal surrounding and placed on a table which has been covered in sack to prevent her from slipping. The doe should be completely relaxed before palpation is attempted. Struggling results in their abdominal muscles becoming tense then making it difficult to distinguishing between foetus and internal organs.

The doe should be restrained by gently holding the fold of the skin behind her ears and over the shoulders. The left hand is placed under the doe's body between the legs in front of the pelvis and

II the fingers of the left on the horns of the uterus. The thumb is

placed to the right of the horns.

Embryos are located by gentle movement of the thumb and fingers in a sideward direction. If small marble shaped object can be felt slipping backward and forward between the thumb and fingers then the existence of the embryo is confirmed. An experienced breeder can determine pregnancy by the 18th — 10th day after mating.

PREPARING FOR KINDLING

I I After 20 — 25 days of gestation the doe's hutch should be cleaned. A newly cleaned and disinfected kindling box containing fresh dry litter

I I such as dry grass, straw or wood shaving, should be placed in the hutch. At about 2gth day the doe will make a nest from the fur that she plucks from her abdomen. This will reveal her teats in preparation for sucking.

Water should be provided the doe liberally at this time to avoid dehydration which may lead to cannibalism. The doe can kindle any moment from now.

KINDLING

The doe will usually kindle or give birth at night. Once kindling has finished the doe will pluck more fur from her abdomen and cover the

I I young rabbits. If the doe did not cover the young rabbits properly add some cotton wool.

When you wish to examine the litter always remove the

doe from the hutch and rub your hand in the damp litter so that they smell of the doe. The young rabbits may not suckle for the first 24 hours, thereafter the doe allows them to suckle once daily. Hence the need to allow the doe and the litter undisturbed for the first few days, since if the disturbance results in a missed suckling, it could be critical for the young rabbit's survival. Do not clean out the kindling box until the rabbits can leave and return without assistance.

When you spot dead rabbits remove them using the method described above. If newly born young are found on the hutch floor they should be returned to the kindling box, again using the method described above.

f The doe vill not return them herself. Rabbits are born without any fur and with their eyes closed. Their eyes open at around 10-11 days and their fur beings to develop at bout the same time. Occasionally bath eyes of the young to allow them to open.

FOSTERING

Some does refuse to suckle their litter, others scatter them. On the other hand a doe may have too many youngsters arid insufficient teats. In any of these cases it may be necessary to foster some or all of the litter into one or more does which kindled on approximately the same day.

Successful fostering is dependent upon the following criteria:

(a) Use foster mother which kindled with approximately 48 hours of the birth of the litter to be fostered. Does which have been kindled longer than this period are more liable to notice the difference between their own and other youngsters and may refuse to accept them.

(b) Where the litter size is too large for successful rearing by one doe the number must be reduced and fostered into another doe if possible within 48 hours.

(c) The attendant's hands must be perfectly clear and free from other animal odours before the nest is examined.

Open the nest from the top only avoiding any disturbance of the side. After gently placing the youngster in the nest carefully replaóe the top of the nest leaving it in the same condition as before.

In commercial rabbitary where many does kindle simultaneously it is a wise policy to carefully examine each preferably on the first day and again on the third day. Any fostering can then be carried out. At the same time check for deformed or dead youngsters which must be killed and burnt.

SYSTEMS OF BREEDING

There are basically two methods of breeding:

(a) Inbreeding and

(b) Out breeding

Another system involves breeding at random within the herd. No direct control is practiced here in terms of selective pairing of does and buck.

(a) Inbreeding:- Inbreeding may be defined as the mating of animals more closely related than the average of the population e.g. mating brother and sister or father and daughter. The main purpose of inbreeding is to increase the probability that the buck and the doe of future generations will be alike genetically in respect of all the characteristic which are required in their progeny. In inbreeding homozygosity is increased. This is due to the multiplication of both recessive and dominant genes. Most undesirable characteristics are dominant and therefore the lowering of quality may result unless the offending stock is quality discarded by very heavy culling resulting in concentration of desirable dominant characters. Undesirable recessives with less noticeable defects may of course be easily overlooked. Economic characteristics which rely on many gene pairs for their expressions depend on inbreeding to establish the traits. No other mating systems are capable of doing this.

Inbreeding is the quickest way available to breeders at the present time of exposing or exhibiting the characteristic make up of their herds. To reduce culling to an economic level the discriminating breeder must be patient and avoid using rabbits which are too closely related. He should mat for instance half brother and half sister in preference to full brother and full sister or father and daughter.

(b) Out breeding:- Outbreeding means mating completely unrelated rabbits of the same breed. This leads to

heterezygosity within the herd and a general breaking down of lines of distinction. Its effects are varied but the tendency is towards the production of progeny which are intermediate between the two parents i.e. they become more uniform.

The breeder who practices this system usually maintains a close herd of does but turns to other breeders for the male line. The aim of this system is to 'upgrade' the stock but the reverse may take place. In view of this fact breeders who wish to practice outbreeding are well advised in the first instance to use a newly purchased buck on only a relatively small number of does. In this way he can determine improvement by carefully recording the resultant progeny.

(c) Linebreeding:- This is a form of inbreeding but differs in that offsprings are kept closely related to a particular ancestor of outstanding merit. This system of mating can prove of great value in those cases where a certain buck has been capable of transmitting outstanding characteristics to his progeny. It is important to mate outstanding individual to their offspring to retain the high quality and not as it is widely practiced to mate to unrelated stock.

GENERAL HOUSING AND EQUIPMENT

Rabbit housing (also known as hutches and cages) and equipment differ from country to country.

The following factors influence the design of hutches and cages:

(a) Availability of raw materials

(b) Temperature

(C) Humidity

(d) Ventilation

(e) Lighting

(f) System of Production

(g) Expertise of the Farmers

No type of housing or equipment is perfect. However care should be taken that some basic requirements are met.

SPACE:- Since the rabbit spends it entire life in the hutch care should be taken that enough space is provided both vertically and horizontally so that the rabbit can carry out its normal activities. Space is very important to ensure good ventilation, temperature control and avoidance of conditions which favour disease outbreaks.

PROTECTION:- Protection is required from injury within the hutch from rain, direct sunlight, wind, sudden noises, predators and human thieves. Convenience and ease of management:- In order for good management hutches should be designed to assist the keeper to carry out activities such as observation, handling, feeding, mating the rabbits, cleaning and disinfection. Because rabbits multiply quickly, planning for expansion should be part of the original hutch designs e.g. hutch may need to be placed on top of another hutch when numbers increase after a few months of production. Males and females from three to four months of age should be kept in separate hutches.

The materials used in construction would usually be those that are available at little or no cost. Interwoven branches, split bamboo, mud, tin, plastic- all can be used with ingenuity. However, whilst these materials may keep rabbits in, they may not keep predators out. If possible, a fence should be built around outdoor hutches and fitted with a padlocked gate. Strength of construction is extremely important.

SITING:- In siting outdoor hutches the following should be considered:-

(a) Place them next to a house wall or fence as this gives shade and protection from sunlight, rain and wind. Too much sunlight may be stressful to the rabbit but too little is also undesirable because the hutch may become damp. There will also be reduced disinfection by sun's ultraviolet rays and vitamin D synthesis by the rabbit may be impaired if it does not experience some direct sunlight. -

(b) In very hot countries it may be important to site hutches under trees or to seek exposed places near water where there are cooling breezes.

(C) Security against thieves and predators is usually best achieved when hutches are next to the keepers house but the rabbits should not be frightened.

Indoor hutches:- Successful rabbit keepers will find that the outdoor hutch system quickly limits production due to lack of flexibility. This can result to overcrowding.

Floor methods of housing:- This method involves keeping the rabbits on the ground, in a fenced area provided with simple boxes for shelter. Several does may be kept in the same area. This method is suited to very dry areas and or where producers cannot obtain or afford the materials to build proper hutches.

EQUIPMENT REQUIREMENTS

A variety of designs and constructions may prove adequate. The basic requirements for each piece of equipment are described below (water and feed troughs).

(a) It should be impossible to tip over

(b) It should discourage scratching out of contents

(c) It should be of adequate size and strength

(d) It must not cause injury to the rabbit

(e) It should be of reasonable cost or be constructed

locally

Forage racks can be fitted inside the hutch, on the outside of the hutch door or in the middle of two linked hutches. They should:

(a) Allow full access to the rabbit and not limit food intake.

(b) Have adequate capacity.

Kindling boxes should:-

(a) Provide a secure, draught proof dry container in which the doe can kindle.

(b) Prevent the rabbits leaving until they are at least 2 — 3

weeks old. '

MEETING THE REQUIREMENTS

Waterers: Waterers can easily be made from tin and bottle. They should be fixed to the hutch wall or floor and if possible, raised from the hutch floor so as not to become fouled with feces or urine. Simple bowls or pots can be used.

Feeders:- Concentrated food is expensive and if used, must not be wasted. Feeders should be fixed to the wall and have an inturned lip and a good depth at least 7-8cm.

Kindling Boxes:- Wood or woven split bamboo may be used. Both closed and open boxes can be successful-30cm high, 40cm long and 30cm wide.

Chapter Six

GRASSCUTTER

(Thryonomys swinderianus and Thryonomys gregorianus)

The grasscutter is found in many forests and savannas of Africa. Its meat, said to resemble suckling pig often, sells far more per kilogram than chicken, beef, pork or lamb. It is the preferred, and perhaps most expensive, meat in West Africa. Indeed in Ivory coast it sells for about $9.00 per kg. With prices like that, grasscutter is a culinary luxury that only the wealthy can afford.

If domestication of this wild species were successful in providing meat at a price similar to that of poultry, (the second most popular meat), markets would be unlimited. However, as production costs are high, long-term research will be required before grasscutter production can be profitable to the small farmer.

In an effort to capitalize on the markets for this delicacy, agricultural extension services of Cameroon,

Ghana, Ivory Coast, Nigeria, Togo and particularly Benin are already encouraging farmers to rear grasscutter as backyard livestock. They furnish breeding stock and information, and maintain central offices for records. In addition, a bilateral cooperation project in Benin has started experimental work on improved breeding methods combined with the study of animal responses under domestication.

In future, this vegetarian animal might becOme the African equivalent of South America's guinea pig, playing an important role in reducing Africa's protein shortage.

AREA OF POTENTIAL USE

The area of potential use of grasscutter are the Humid and Sub-humid Africa South of the Sahara.

APPEARANCE AND SIZE

Grasscutters are robust animals with short tails, small ears, and stocky bodies. Taxonomically, they are more closely related to porcupines than to common rats or mice.

Although many varieties have been described, there are probably only two species. The larger (Thryonomys swinderianus) weigh.s 9kg or more and has a head and body length of up to 60cm. The smaller species

52

I

(Thryonomy gregorianus) may occasionally reach 8kg and a body of length of 50cm.

Both species have yellow-brown to grey-brown bodies with whitish belies. The fur is extremely coarse firm, and bristly — reflecting the animals kinship to the porcupine. The tail is scaly and has short, sparse hairs.

Both species have thick, heavy claws and enormous orange incisors that can chew through even the toughest vegetation. They tear holes in corrugated iron fences. Nevertheless, they do not bite when handled, although their claws sometimes cause injuries.

DISTRIBUTION

Grasscutters occur in grasslands or in wooded savanna throughout the humid and sub humid areas of Africa South of the Sahara. They often live in forest — Savanna habitats where grass is present. They do not inhabit rain forest, dry scrub or desert, but they have colonized the road boarders in forest regions. Distribution is determined by availability of adequate or preferred grass species for food. Specifically, Thryonomys swinderianus occurs in virtually all countries of West, East, and Southern Africa.

Thryonomys gregorianus occurs in savannas in Cameroon, Central Africa Republic, Rep. of Congo, Sudan, Ethiopia, Kenya, Uganda, Tanzania, Malawi, Zambia, Zimbabwe and Mozambique.

STATUS

Despite heavy hunting, these animals are not threatened with extinction.

Nonetheless, many individual populations are well below carrying capacity, or are extinct because of local over exploitation.

HABITAT AND ENVIRONMENT

PThe large grasscutter (T. swinderianus) generally lives in swampy, low- lying areas, especially along river banks and the boarders of lakes and streams. Occasionally, it is found on higher ground among bushes and rocks, living where savanna grasses are dense and tangled enough to afford good cover. In Ivory Coast and Southern Guinea, for instance, grasscutters are found (and hunted) throughout the savanna zones. And they can occur in close proximity to farmlands and people as in southern Nigeria.

BIOLOGY

Although the precise diet in the wild has not been determined, grasscutters are vegetarians. They consume nuts, barks, and the soft

53

I

parts of grasses and shrubs. They particularly favour the elephant grass and sweet potatoes. They commonly "raid" casava and yam plantations and are considered local pests.

Grasscutters reproduce year-round, although the births seem to peak at certain times of the year, correlated with weather conditions. Probably one male takes several females, and the family group possibly has more than one generation of young. The gestation period is about 152 days. Apparently, litters normally contain between 2 and 4 young s, but in Benin and Togo some litters of up to 11 or 12 are reported. Newborns are fully developed, their eyes are open, they weigh approximately 80g, have thick fur, and quickly become accomplished runners.

BEHAVIOR

Although they commonly forage in groups, grasscutters are generally solitary. They are nocturnal and they travel at night through trails in reeds and grass, often to water. Most specimens are males, possibly because males lead the groups and are most prone to whatever form of attack.

When alarmed, these micro ruminants stamp their hind feet and give a strange booming grunt. When fleeing, they can run very fast and, given a chance, will take to water for they swim with ease.

For shelter, grasscutters usually weave nests of matted vegetation or scoop out shallow burrows.

USES

In a broad geographic band across sub Saharan Africa, cattle raising is severely limited by trypanosomiasis. Here, other sources of protein, including rodents, are traditionally used. Thus, grasscutter meat constitutes an important food fOr many Africans. The animals are mostly caught and prepared for family use, but some are sold in markets and especially in roadside stalls. Many families depend exclusively on selling bush meat, particularly that of grasscutters. In Accra, Ghana, during one year, 73 tons of grasscutter meat was sold in local market. This represents more than 15,000 animals. In Southern Afriba, too, people find that these rodents make tasty food, although they may cut off the tail to make the carcass look less rat — like.

54

The meat is usually eaten smoked, and is so much in

demand that grasscutters are hunted in organized drives with spears, dogs, guns and sometimes fire. It is considered excellent, especially when cooked in soups and stews or barbecued. It has been described as resembling venison in flavour, but it is dark like the meat of wild duck.

HUSBANDRY

In the savanna area of West Africa, people have traditionally captured wild grasscutter and raised them at home. As an extension of this, organized grasscutter husbandry has been initiated in West Africa. The animals are provided with marshy, tightly fenced areas with plenty of plant cover. The young are harvested from these and raised separately.

Ghanaian researcher Emmanuel Asibey, a pioneer of this research, reports success at getting such captive stock to reproduce. To this end, farmers are provided with boxes and foundation grasscutter colonies. They are taught how to rear and feed the animals for home consumption or for cash income. Basically, the farmers make available large sheds where the animals can move freely. The farmers also provide piles of grass, sugarcane and other foods. A grassscutter reportedly takes about a month to adjust to such confinement. High mortality can occur in this period. The average dressed carcass is 64 percent.

The Wildlife Domestication Unit of University of Ibadan, another pioneer of rodent domestication, has also reported the potential of domesticated grasscutter colonies.

In Benin Republic research on grasscutter breeding and husbandry and feeding is also being implemented by the Ministry of Rural Development.

LIMITATION

Grasscutters can devastate such crops as rice, sugarcane, soyabeans, peanuts, yams, cassava, sweet potatoes, oil palm seedlings, maize, young rubber, sorghum, wheat, and telferia. Therefore, as with most rodents, they should be reared only in areas where they already exist.

In past years, captive animals in Benin have suffered fatal Clostridium infections during September and October. In 1986, a broadspectrum antibiotics was given with outstanding result. During this season, the animals also suffered from ascarid worms, which were also successfully treated with standard drugs.

OTHER NAMES OF WHICH GRASSCUTTER ARE CALLED

Thryonornys swinderianus could also be called

(a) Marsh cane — rat

(b) Larger cane rat

(c) cutting grass

(d) Aulacode

(e) Agouti

(f) Sibese (Guinea)

(g) Oya (Yoruba)

(h) Simbiliki (Congo)

(i) Ndesi (Swaheli)

(j) Ivondue (South — East Africa)

(k) Nchi (Ibo)

POSITION OF GRASSCUTTER IN THE ZOOLOGICAL SYSTEM

Table

Strain:

Chordate

Sub strain:

Vertebrata

Class:

Mammalian

Order:

Rodentia

Sub order:

Hystricomorpha (porcupine relatives)

Supre family:

Petromuroidea (rock-rat like)

Family:

Th ryonomyida (grasscutter)

Genus:

Thryonomys fitzinger 1867

Species:

Thryonomys swinderianus Temminek 1827 larger grasscutter,

Thryonomys gregorianus Thomas 1894 lesser grasscutter

Chapter Seven

POULTRY

Exotic Birds Production

Definition: Poultry in agriculture generally refers to all domesticated birds kept for eggs or meat production. They include such birds as fowls, ducks, geese, turkeys, pheasants, guinea fowl, ostrich etc.

IMPORTANCE OF POULTRY INDUSTRY

1. Meat Production (2) Egg Production (3) Feather (4) Income

(5) Manure (6) Hobby (7) Industrial uses:

Others uses include:

(a) Vaccines

(b) Pharmaceuticals,

II (c) Paint,

(d) Vanishes,

(e) Adhesives,

I I (f) Printer's ink from eggwhite

(g) Soap, Shampoos from yolk

(h) Shell in mineral mixture and for fertilizers

(i) Use of chicken feathers for making pillows, mattresses and

insulation materials.

ADVANTAGES OF POULTRY FARMING

1. Adapted to small environment and space

2. Prolific and productive -maturation period short

3. Large scale production creates employment

4. Less acutely affected by unfavourable weather condition

5. Availability of improved breeds

6. Car' be an industry and profitable enterprises

7. Waste products converted in to poultry feeds for the production of high quality protein eg Eggs and meat.

8. Quick income return e.g broiler for only 7 weeks

9. Income is distributed throughout the year

10. Can be full time or part time

11. It is a broad field of science involving knowledge and proven practices as much as explanation e.g. feeding, breeding, management and marketing, medicine, engineering, technology etó.

12. Helps in fighting animal -based protein deficiency in Nigeria.

13. Itischeap

57

I

14. Women, children, elderly, literate and non-literate can be successful in poultry production as long asthey have the technical know.

MANAGEMENT OF DAY OLD TO ADULT CHICKS

The above involves brooding of day old chickens. It is the care of the chick from a day to six weeks of age. It consists primarily of the provision of heat, air, water and feed. The efficient combination of these factors determines the level of physical and physiological development and mortality and morbidity of the chicks.

On arrival of the chicks, they should be unboxed and inspected individually with clean hands before placing them under the brooder. The brooder snould have been set -water, feed, heaters. Use shallow trays for feeding. A strong disinfectant should have been used to disinfect the house two days before the arrival of the birds.

Rearing: Rearing is the care from 7 weeks to point of lay of the domestic fowl or care of growers. This is to ensure that the development of the fowl during this stage is satisfactory. Increased ventilation, adequate quantity of feed, (35-45g/bird/day), no vaccination at this stage as vaccination should have been completed at 6 weeks, managed either in deep litter or cages.

Feedinenerally on /Bird I Day Basis: 0-3wks 7-hg, 4-5wks 25- 45g, 6-735-45g, 8- llwks 40-60g, 12-15wks 45-70g, 16-17wks 60-90g,

18-1 9wk 54-77g, 20-22wks, 68- 110g. But we may not have the time and facility to measure these things so the rule of the tumb here is to feed the birds ad-lib or at that adult age 25kg (a bag of feed) to 300 birds *Note

MANAGEMENT OF ADULT BIRDS

The adult fowl is a layer, a breeder or full-grown broiler. From point of lay, the females undergo a pattern of physical and physiological changes manifested partly in the frequency of egg production, egg size, body size, comb colour and efficiency of egg production. Based on these changes the pullet years (first year in lay) could be divided into three phases.

(1) Peak Period -Egg laying increases from nil to peak which occurs at about 2-3 months after point of lay. Egg size also increases at about forty —second week of age.

58

(2) Second Phase -Gradual fall occUrs in production but not in egg size and body weight. Lasts till sixty-secoOd week. Egg production drops to about 65%.

(3) Third Phase -Remaining pullet year and terminates in moulting. Conspicuous drop to practically nil in egg production. Egg size does not decrease. Efficiency In rate of feed conversion declines.

Note:- In the tropics 180 -200 eggs per year is regarded ad adequate though it could be more.

Provision of clean and enough water and fresh feed are very necessary for the over all health of the birds.

ROUTINE MANAGEMENT OPERATIONS

1. Remove dead birds to prevent contamination

2. Fresh feed added to stale feed in the trough after litter and droppings in the (!1ter have been removed.

Mix them with finger.

3. Water fountain or troughs are moved out and thoroughly cleaned and replenished with fresh water.

4. First collection of eggs occurs at 9.00am followed by 12noon to 1 pm and lastly 4pm to 5pm.

5. Turn litter frequently after collecting eggs

6. Feed ad lib.

OCCASIONAL MANAGEMENT OPERATION

1. Culling -Due to non-lays, deformity, vice habits, poor production or poor egg quality, poor growth rate, cull every unthrifty bird. Watch the comb, pelvic girdle and the abdomen and cull accordingly.

TABLE BIRD PRODUCTION

The production of table birds has grown into a major industry in many advanced Dountries. However, it is a more specialized industry than egg production.

Poultry meat used to be derived predominantly from spent layers (boilers). This is still the main and the preferred source of poultry meat in the developing countries although there is an increasing shift to broilers.

These are fast growing birds, which reach market

weight of 1.6kg to 2.0kg in 8-12 weeks.

FACTORS OF ECONOMIC INTEREST IN BROILER PRODUCTION.

1. Market weight attainment of about 1.6- 2-0 kg in 8 weeks or about

12 weeks.

2. Good body conformation

3. High efficiency of feed utilization

4. Low mortality.

STRAINS OF BROILERS

Commercial broiler chicks are derived from the Plymouth Rock, Cornish, Sussex, Rhode Island Red and White Leghorn. The commercial chicks

are mostly multiple crossbreeds.

ENVIRONMENTAL FACTORS FOR BROILER AND THEIR MANAGEMENT

The essential environmental conditions for broilers and pullets are similar. However, broilers require higher levels of certain nutrients like protein and energy. They also require higher levels of starting temperature of about 35°C.

1. Good ventilation necessary

2. Reared traditionally on deep litter

3. Mass brooding

4. No too much lighting required. Enough to see feed

5. Immunize your birds.

TYPES OF BROILERS

1. Roasters- Usually, males of dual- purpose hybrids

2. Poussins -By product of hatcheries

3. Capons -Simple or chemical treatment to remove the action of male hormones or increase the action of female hormones. The testes of the males may be surgically removed while both males and females may be given hormones preparations (estrogen). Usually only the males are caponized.

HYBREED BIRDS FOUND IN NIGERIA AND COUNTRY OF ORIGIN

(1) Layer Type Country of Origin

Hyline Britain

Harco Britain

Thorn bar 909 Britain

Yarkon Israel

Shaver Star Cross Israel

Low Mah Israel

0

V

(2) Broiler Type

Cobb Bhtain

Anak Israel

India River USA

Jupiter USA

Bitch USA

COCKEREL

Literally meaning young cock. It is a by-product of breeding operation. Takes time to grow. Gestation period up to six months. Reared the same way as pullets. Under normal circumstances should not be reared but Nigerians prefer cockerel to even broilers. So it is a big business in Nigeria. Itinerant poultry seller goes on bicycle to distribute to housewives and other prospective buyers in what is termed 'brood and sell'.

BROODING PROPER

The sides of the houses should be covered with cellophane or mats or sacks. Maximum ventilation is restored from about the 5th week of brooding the birds. Avoid marching while inside the brooder pen. Use appropriate feed.

Use kerosene stoves, good hurricane lamps and coal pots for heating the brooder. Never you depend on electricity. Do not travel faraway from the brooder house. Use sugar in lukewarm water solution for the first one-week. Allow them drink for the first one hour before introducing feed to rehydrate them.

SPECIAL OPERATIONS (S P 0)

1. Debeaking

2 Dubbing

Dewarming

4 Vaccination

5 Despurring

VACCINATION, MEDICATION AND MANAGEMENT PROGRAMME FOR RAISING BIRDS"

Table

4th Day

Ndv ,, antibiotic/Vitamins

2' week

3rd week

vitamin Administration, Gumboro 1st Dose

Vitamin Admin. 1st Dose NDV-L, Coccidiostat starts till the 8 week

4Th week

Refresh the litter, Gumoboro 2nd Dose

5Th week

—

Change feed to broiler finisher if broiler, Vitamin admin, NDV-L 2' Dose

6th week

7th week

Vitamin Administration, Gunboro 3id Dose Refresh

the litter, Dewarm

8 week

Maintain your good management, change to growers feed.

h week

Continue till they drop the first egg before changing to layers. FTV (Fowl typhoid vaccine)

i0" week

FPV Fowl Pox Vaccine

12 week

N DV — Komorov st Dose

16th week

NDV — Komorov 2 Dose -

18th week

Debeakng

BROILER

Day 4

Give Vitamin, NDV -

2 week

Give Vitamin, Gumboro 1st Dose

3rd week

Give Vitamin, NDV-L jst Dose

4th week

Give Vitamin, Gumboro 2nd Dose

5th week

Give Vitamin, NDV-L bose change feed to broiler finisher -

6th week

Give Vitamin, Gumboro 3 Dose

7th week

Arrangement for Marketing

— 12th week

Arrangement to dispose the birds.

POULTRY DISEASE

Poultry diseases are a major cause of financial losses in the poultry

industry. Loss may arise from deaths or poor levels of performance.

Some of the major poultry diseases are listed below.

VIRAL DISEASES OF POULTRY

1. New Castle Diseases -Caused by Paramyxo virus. The virus is shed during incubating stages throughout the clinical stage,

present in exhaled air, respiratory discharge etc. Symptoms are respiratory and nervous. Control by vaccination

2. Gumboro -Caused by infectious bursal virus. Affects only growing chicks Control is by isolation and vaccination. Symptoms are anorexia, depression and ataxia (un-coordination) and finally death.

3. Infectious Bronchitis: Respiratory tract infection in birds of all ages. Caused by an RNA virus with many strains. Prevention is by good management and hygiene. No known treatment.

4. Infectious Laryngo Tracheitis: Virus disease of domestic fowl and pheasants affecting respiratory tract and mainly of adult and young birds and characterized by respiratory distress, coughing, rattling and lacrimation. It is caused by a DNA virus. Control is by keeping the birds quiet, lower dust, use mild expectorant and vaccinate all adults during outbreak

5. Fowl Pox: Mainly in chicken and turkey and

characterized by formation of nodules on the skin and progresses to form heavy scabs and membranes in upper digestive and respiratory tract. DNA virus causes it. Treat with palm oil or iodine.

6. Avian Encephalomyelitis (Epidemic Tremor) - Causes partial paralysis, ataxia and tremor. Control is by good housing and hygiene and avoidance of fecal contamination.

7. Neoplastic Diseases (Aviain Leukosis Complex): Signs are pale and shrunken combs. Enlarged liver felt. Confirmation under post mortem shows lesions in the bursa of fabricus, spleen, liver, bone marrow.

BACTERIAL DISEASES OF POULTRY

1. Solmonellosis:- A group of name of the diseases caused by members of the genius selmonellae characterized by acute course involving a generalized infection of the body organ or a chromic course involving localized infection. There are three members of the salmonellosis group.

(a) Pullorum Disease (b) Fowl Typhoid (c) Paratyphoid.

2. Pasteurellosis: Chronic or acute generalized or localized infectious disease of adult birds and characterized by sudden onset, high mobidity and mortality.

3. Tuberculosis: A slow spreading, usually chronic infectious contagious infection of birds usually in mature birds and characterized by gradual loss of weight, unthriftiness and emaciation caused by mycobacterium species.

4. Mycoplasmosis (CRD): Characterized by rales and coughing. Caused by mycoplasma galisepticum. Tylosin used for treatment. Streptomycin and other antibiotics could be used not penicillin.

5. Infectious coryza: An acute respiratory disease of adult birds (2-5 months) and characterized by sneezing, rales, coughing and marked swelling of the face. It is caused by the bacterium Haemophilus gallinerum. Treat with streptomycin, Erythromycin and sulfonamides.

6. Bluecomb Disease (Avian monocytosis, pullet disease). It is an acute or sub-acute condition primarily of young layers characterized by high morbidity, low mortality and sudden onset. Cause of the disease not yet established. Signs are foul smelling diarrhea, wilting and blueness of comb and wattles. Treatment is by potassium in form of malasses in ration and tetracycline group of drugs.

DISEASES CAUSED BY FUNGI

1. Aspergitlosis (Brooder Pneumonia)

2. Thrash (moniliasisis/candidasis

3. Favus (whitecomb).

4. Mycotoxicosis

Most important here is Aspargillosis caused by Aspergillus 4avus. Sources of infection are mouldy feed and moist litter. Treatment is by use of copper sulphate solution in drinking water, Grissofulvin and Nystatin. Signs include dyspnea, gasping, accelerated breathing, nervous involvement and emaciation.

DISEASES CAUSED BY PROTOZOA

(1) Coccidoisis

(2) Trichomoniasis (inflamation of the vagina (urethra in males) caused by Trichosoma vaginalis.

(3) Histomoniasis (Black head) ",

(4) Leucocytozoon (Leucocytolysis) disease

(5) Malaria

The most important here is coccidiosis. The causative agent is the genius Eimeria which has eight species of protozoa. Two of them cause wasting diseases.

(a) Elmeria teneilla (caecum), Eimeria acervulina (duodenum). Others less pathogenic or non pathogenic at all are E.necatrix E.brunetti,

E. Mitis, E. magna.

Processof infection: The non-infectious unsporulated oocyst develops into a sporulated oocyst after one or two days or several days of warmth

and moisture in the ground or litter depending on species. The spores hatch in the intestine (sporozoites), invade the intestinal lining, multiply (shizogony) and gametes (merozoites) are produced.

NIB Coccidia are almost universally present in poultry raising operations, but clinical diseases occur only after ingestion of relatively large number of sporulated oocyst.

CONTROL

(1) Use of sulfonamide drugs

(2) Amprolium

(3) Embazin Forte

(4) Nitrofurans

(5) Vaccination

HELMINTH PARASITES

1. Nernatodes (Round Warm)

2. Cestodes (Flat Wcrms)

3. Trematodes (Flukes)

EFFECTS ON POULTRY

1. Retardation of growth rate

2. Reduced rate of laying

3. Anaemia

4. Loss of weight

5. Enteritis

TREATMENT

1. Piperazine Salt fl \i ematodes

2. Thiabendezole

3. Diachlorophan

4. Dibutynorate Cestodes

ECTOPARASITE DISEASES

1. Ticks

2. Lice

3. Fleas

4. Mites

5. Mosquitoes

6. Other biting flies

TREATMENT

Organophosphate preparation

A. Arsontol

B. Carbaryl

C. Servin

D. Gamalin

E. DDT

F. Izal

FEEDS AND FEEDING

There are various commercial feeds in the market for the various types of poultry.

a. Broiler starter

b. Broiler Finisher

c. Chicks mash

d. Growers mash

e. Breeders mash

f. Layers mash

You have to watch out for inferior feeds. Insist on buying superior feeds.. I recommend Guinea feeds and Sanders feeds. There may be other good ones. The feeding of the birds has already been discussed.

ALTERNATIVE OF CONVENTIONAL FEEDS

Poultry like humans have feed requirement because of the monogastric nature of their digestive system. These requirements must be met if we are to get the desired result.

However, administration of unconventional feeds could be applied at a minimal level. In this direction therefore fish remnants, waterleaf, Centro and some known succulent grasses could be fed to the birds as supplements. The green leaves should be thoroughly washed to avoid introduction of diseases into the poultry.

EQUIPMENTS

Rakes, buckets, spades, cutlasses both chicks and adult drinkers, chicks and adult feeders, headpan, scissors, hand gloves e.t.c.

GUIDES TO GOOD MANAGEMENT

It is said that in poultry production, prevention is better and cheaper than cure. For this reason the following guidelines should be employed:

(1) As much as possible, use day-old chicks as your starter instead of point-of-lay fowls or growers.

(2) The farm should be located in a favourable geographical site in respect to other poultry farms

(3) Proper location of the building in relation to prevailing wind dire Dtion or current.

(4) Long range planning and programming of movement patterns of various vehicles and equipment in and out of farm.

(5) Farms should not be too close to oneanother.

(6) Avoid mixing birds of different ages.

(7) Poultry houses to be constructed in such a way as to exclude wild birds, rodents and reptiles.

(8) Provide foot dips or permanent foot wears just for the pen.

(9) Adequate ventilation to be provided to control dust and ammonia build up.

(10) All surfaces inside to be made of impervious

materials to permit thorough washing.

(11) Adequate facilities should be provided for occasional feed and water medication.

(12) Rats, mice and other rodents should be kept out of the feed and water.

(13) Visitors and customers should not be allowed into the houses. If they must be allowed then they will be clothed with farm coats and then dip their feet.

(14) Carcass disposal should be very adequate, as dead carcass often constitute a source of infection. It should not be eaten nor sold.

(15) Regular cleaning and dis-infection of the building and premises.

(16) Do not feed your birds on mouldy, stale or poor quality feed.

(17) Do not allow your litter to stay too long.

(18) Always have a foot dip containing strong disinfectant at the entrance to the poultry house.

(19) All-in all-out method should be used.

HOUSING AND MANAGEMENT

Management System Housing

a Extensive System I Range System

II Fold Unit System

b Intensive System I Deep Litter System

II Wire and Slatted Floor System

Ill Straw yard System

IV Battery Cage System

c Semi-intensive System I Typical semi-intensive unit

II Straw Yard

The intensive system requires solid structures. The roof may be full span (gable) or lean to -type. The height of the ridge is related to the width of

the house; the difference between the ridge and the eave height is about onequarter of the width of the houses of 7.2m or less in width. Wider houses have a lower ratio of ridge height to width, so that the house will not be too high. Lean to roofs may create the problem of rain drifting into the house. With all types of roofs, the eaves should be long enough to check rain.

RECORD KEEPING

It is essential to keep daily, weekly, monthly and yearly records for all classes of poultry. The records provide information on the trend in profit margin when the business is in progress and data for assessing profitability when the business is completed; they are therefore essential when making business decisions. Records provide information that can be used to detect and correct management errors, and to detect and stop the spread of diseases. The records may also be needed for governmental planning on a national basis.

There are no hard and fast rules as to how the poultry records should be kept. Nor is there any particular format the record should take.

The different types of records should include records of stock or of mortality, feed consumption, egg production and egg quality and sales. There may also be records of veterinary services. The format below could be of help and a guide:

Table: 7.2

Date culls Dead Sales Balance SPO Feed Remarks

NIB SPO Special Operation

Disease outbreak could be indicated under REMARKS

Chapter Eight

POULTRY

DOMESTIC DUCKS PRODUCTION

Introduction

In the Nigerian context domestic ducks can be raised on any type of condition with little or no attention. These birds are more rugged than the indigenous fowls, lay more and larger egg, grow to a greater size, require little in the way of housing and are not so susceptible to disease and parasite. They are also better able to protect themselves from marauders and are excellent foragers. Commercial, large scale duck keeping is now expanding outside the South-East Asian tropic where they are the sole source of livelihood of a considerable number of people (Warren 1972).

ORIGIN

Most breeds of ducks appear to have been derived

from the wild mallard (Anan Platyrhyncha). It is likely that the centre of domestication was somewhere in South-East Asia. Zeuner (1963) has suggestd. t Western Asia might have been a centre. Moscovy breed has beërom the American tree duck (Cairina Moschata).

BREEDS

The breeds in Nigeria are kept for their meat. In Nigeria the eggs are seldom eaten due mainly to the colour and size of the eggs but in Asia the reverse is the case. The major breeds are:

a. The Moscovy

b. Indian Runner

c. Mageswari of India

d. The Chinese of Indo China

e. The Java of Malaysia and Indonesia

f. Khaki Campbell

g. Rouen ducks

h. White Pekin

Aylesbury

j. Pintail

k. Shovellar

Under subsistence production ducks of the breed of Indian Runner, Mageswari and Chinese of Indo-China and Java of Malaysia and Indonesia achieve mature body weights of approximately 1.8kg and lay

70

A

on the arrange 60-80 eggs per annum. But under intensive, improved husbandry condition Indian Runner drakes (male) will weigh 2.0kg and the ducks (hen) lay 200 eggs annually.

BREEDING AND INCUBATION

Breeding of ducks in Nigeria has not developed to the level of using artificial incubators. For this reason no steps have so far been taken to house then and properly feed them. Under this extensive system being used today much can still be done to improve on their performance.

Normally these birds are allowed to free-range and mate at random. The belief that effective copulation can only take place in water is mistaken, although it is desirable that ducks should have access to water on which to swim. With this they are able to keep themselves clean and nesting ducks are better able to keep their eggs at the correct humidity.

The breeds in Nigeria are more or less a mongrel stock and so much variation in terms of production is evident. Some breeds hatch up 20 eggs and some 10-15 eggs. Muscovy ducks can incubate their eggs satisfactory and are capable of hatching up to 30. In which ever breed warm water should be lightly sprinkled on the eggs daily (tepid water).

Incubation period is 4 weeks for all breeds except Muscovy, which is 5 weeks. The mature period for ducks to begin to lay egg is 6 months.

In countries where ducks are produced in large commercial farm the eggs are artificially incubated, using temperature of 0.6°C (1°F) lower than is needed for hen's eggs. Relative humidity should never face below 60 per cent. After hatching it is simpler to sex duckling (young ducks) than in chickens.

REARING

Duckling should be transferred to a brooder as soon they have dried in the case Df artificial incubation. Temperature of 29.4°C (85° F) is ideal during the first few nights which could declined to 18.3°c (65° F) within 14 days. 'Cold' type of brooder is perfectly satisfactory in the tropics.

Ifducklings have been incubated natuiUy (using the ducks) they could be left with their mother for about 4 weeks. Most breeds are good

mothers.

SYSTEMS OF PRODUCTION

The management method used for producing fowls are also used for ducks depending on interest and market situation.

In Nigeria what I may call the 'Anwuri System' is best for producing the maximum number of ducks with little effort and maximum results. This method involves riding about for ducks with a drake on a free range and when the ducks hatch they are allowed to remain with their mother for about three weeks after which they are separated and their ducki housed and fed and watered separately properly. Deep plast containers are filled half way with clean water to allow them swim when they desire. Green leaves could be occasionally given them. Different pens could be created for different ages of the ducklings that may be expected. While outside the ducks and the drakes and the ducklings will be provided with addition feeds and water. Water particularly makes them to stay within the area. Within 10 weeks the ducklings will attain table weight and those that will be retained for breeding purposes will be retained while the rest are sold off. Ducks grow faster than even broilers especially when they are properly fed. In all cases what should be allowed to roam or free range is the breeding stock. In some cases the ducks hatch up to 14 duckling and adequate arrangements should be made to ensure the ducklings are not harmed within the first 3 weeks. They should be allowed to remain with their mother. Flood water, reptiles, rodents, kites and hawks are major source of loss within these 3 weeks.

DUCKS MEAT

Ducks' meat is a wholesome meat. It is even richer than the meat of fowl. The reason many do not savour the meat of duck is the method of preparation. Ducks' meat is naturally a broiling type and should therefore be broiled (deep Frying) and not eaten after straight boiling. The earlier people make up their minds about this cheap and easy to produce source of rich protein the better for us. Ducks' meat like any other type of poultry meat is low in cholesterol and better for the human body than beef chevon and mutton.

DUCKS EGGS

Ducks eggs like the eggs of hens, turkey, guinea fowl and pheasants are good for the body. The obsession caused by the size and colour of the duck's egg should not be. It is merely artificial. Ducks are naturally water birds and clean. It is when there is no water to swim in that they go far anything-mud, stagnant water etc. The egg could be fried or boiled with a pinch of salt. The duck is not a native of Africa and so continuous efforts should be made to understand and appreciate the value and relevance of this all important animal. You cannot just hate what you have not tried before. A trial will convince you.

DISEASES

Ducks are generally resistant to stress and clearly exhibit a tolerance to most poultry diseases, consequently, they do not require routine vaccination. Apart from preying and traumatic injuries inflicted by snakes and man ducks hardly die. However, they suffer from cholera and a form

of hepatitis. They are also susceptible to and intolerant of contaminated and moldy feed and also deficiency of manganese in their diet. Liver flukes and internal parasite could be a problem.

FEEDING

In our Nigeria situation, you can feed the duckling with chicks mash to start them but finish then using grower ration or broiler finisher if you can afford. Add supplementary feeds such as succulent legumes, grasses and agriculture by-products and kitchen waste as earlier discussed.

RELEVANCE OF DUCKS OR ADVANTAGES

a. Ducks have more meat per unit area than the domestic fowl

b. They grow faster with minimum feed requirement.

c. When properly handle they are cleaner animals than domestic fowls

d. Too much housing not required in ducks production

e. The eggs are bigger

f. They are more resistant to diseases and stress condition

g. They are better foragers than domestic fowls

h. They adapt more easily than the domestic fowls

i. The meat has better texture than those of other poultry meat

j. They are friendlier than domestic fowls

k. Their ability to fly and remain air-horn for minutes and pearch on taller trees keeps them away from danger (preying animals) and other animals

I. They are not noisy like the domestic fowls•

m. They do not scratch the ground and scatter refuse as do the domestic fowl there fore are environmental v=•

n. They are amphibious birds and this becorres a'd'vantageous in flooded areas.

o. They multiply faster than domestic fowls and geese

p. They produce rich manure

q. The eggs are useful for industrial uses such as vaccine and paints etc.

r. The manure could be used for fertilizing fish ponds

s. Ducks production could be exciting past times.

SUMMARY

Protein deficiency in Nigeria is a serious issue whether we believe it or no. One of the ways to fight this ugly situation is to invest in duck production. It will require little capital, little space and little effort. You can start today.

HENRY OJO

Chapter Nine

POULTRY

Production Of Local Chicken Or Indigenous Birds

The local chicken is also included in the list of domestic fowl, of the order Galiformes and family phasianidae.

These birds are cherished by Africans because of the peculiar texture of the meat and its juicy palate. Unfortunately these birds are hard to come by and when they are available are costly and sometimes not worth the price because of the smallness of the size, the juiciness notwithstanding. This write up is out to look into the possibility of improving on the existing methods of producing these local birds which have obvious potentialities of meeting or improving the present poor level protein in take of our rural populace.

We should note that no deliberate effort is ever made to feed or water the local birds. Even at the worst of dry season nobody ever makes any effort to provide water for the birds. They wade through the annual dry spells with little or no water at all. They scratch on human feces. Feeding is just from the garbage dumps and forage crops if any. Even when they have genetic potentialities to do well, poor nutrition and environment inhibit their development. They therefore cannot express to the limit that nature ha:s set for them. In addition to the above mentioned problems these birds are left to the mercy of rodents, reptiles and preying birds such that over 80% or more of the chicks they hatch do not survive. In some cases they are left with nothing at all.

They spend most of their energy searching for food and water. These energies could have been utilized in tissue synthesis and improved egg quality or adipose tissue formation.

WAYS TO IMPROVE ON LOCAL CHICKEN PRODUCTION

Free ranging could still be used but ready-made commercial feed and water should be provided for the birds first thing in the morning before they go searching for food.

ii. Few days after hatching they should be placed in a cage with slatted floor where they will not come in contact with their droppings. The cage could be moved from time to time.

iii. Water and feed could be provided for them there in the cage using appropriate feeders and drinkers. Start with Broiler starter.

iv. Feeds could be compounded locally using some cereals, cassava dried and crushed, beans remnants, fish remnants, corm chaffs,

maggots, green leaves, and other kitchen wastes.

v. The birds will mature in less than 5 months as against 8 months or more if they were roaming.

vi. The cages could be partitioned to accommodate the hen and her chicks only. Add grit to their feed. Ashes and soot from the hearth should be introduced to them as a source of mineral.

vii. They should be shielded from rains and sun and also kept in a secure place where reptiles and rodents will not have access to the birds.

viii. Arrangements could be made to dispose of mature birds to make room for the hens that have hatched.

ix. When the chicks have feathered properly their mother could be released from the cage to continue in the next round of breeding.

x. Occasionally vitamin powder could be added to the drinking water and vaccine also administered for required improvement. Dewarn after every 3 months.

xi. Slow growth and unthriftness are due to the absence of the above requirements.

xii. Improved breeds could be introduced to the local stock to improve their genetic potentials and thereafter selection could be introduced to get rids of unthrifty birds.

RELEVANCE OF LOCAL CHICKEN

i. They produce beautiful feathers than exotic breeds

ii. They are resistant to disease than exotic birds.

iii. The texture of their meat appeals to the African more than the exotic birds.

iv. The meat is also juicier

v. They are stronger fighters

vi. Local birds and their eggs are both used for sacrifices and never the exotic birds. They appeal to the gods. Their egg and meat are more appealing to the gods of Africa

vii. They are good time-keepers

viii. They are god pest controllers

ix. Their eggs though smaller are rich and sweet

x. They still remain a source of the stock for a thoroughly adaptable and indigenous breed for our condition.

xi. Grown up cocks are alarm raisers.

xii. Their beautiful feathers, serves as ornamentals for our environment.

xiii. From my research they are easily adaptable and show little or no cage fatigue at all. They could be a good source of table eggs supply when their potentials are properly harnessed.

xiv. In rural communities these birds often serve as ransomes for offences committed by their owners.

xv. They are used for rituals and other traditional ceremonies in Africa

xvi. In some communities they are kept as pets.

xvii. Their meat is a delicacy in making pepper soup.

With these in mind we can see that we cannot afford t loose or neglect the indigenous or local chicken especially with the high cost of the improved breeds. Every person should at the least keep two hens and a cock and apply the recommended improved practices. You find out that within six months you could be enjoying chicken and eggs at your pleasure.

THE DANGER OF CONSUMING LOCAL BIRDS REARED EXTENSIVELY

1. Because local birds are more or less scavengers they are worm- infested and could be a source of worm infestations to consumers. Many species of worms could be contracted from the consumption of local birds-Nemetoodes (round worm), cestodes (Flat worm) and Trematodes (Flukes)

2. Recent research has also proved that significant quantities of lead (Pb) are found in animals reared extensively.

3. Ectoparasite infestation is also prevalent in birds reared extensively. Because they roam about they could contract pests from far and wide and eventually roost within human habitation.

4. Diseases of Zoonotic significance could occur since there is no monitoring such diseases as Tuberculosis and Salmonellosis:

I. Pullorum Disease

ii. Fowl Typhoid [cannot be ruled out.

iii. Paratyphoid

Chapter Ten

SNAIL PRODUCTION

There are more than thirteen known species of snails. The most important are the Achatina species and the hn species. Others are the Helix species, Littoria species, strombus species, Ecma species, otala species, vitrinia species, Eremma species, Megalatrectus species, Ampullarius species and Limicolaria species.

Ampullarius species is a common fresh water snail found in Akwa Ibom areas of Nigeria and is harvested during the rainy season. Achatina and Arachachatina are the two giant land species of snails found in West Africa coastal countries like Nigeria and Ghana. These species are recommended for commercial farming.

Presently most snails seen in the market are picked from bush, either at night or during the day. Sometimes bushes are set on fire to facilitate snail hunting. Like any other business, there is need for an organized form of production of snail to ensure a steady supply. Only few organized snaileries exist in West Africa.

From all perspective, snail farming on large scale is relatively cheap in terms of pace, building materials, money and time required for routine work and for maintenance when compared with poultry or other livestock enterprises. Without much ado your snail farm can start at a small corner of your compound as is done in poultry and other livestock enterprise.

SNAIL AS FOOD FOR MAN

It is not as if snail is a new food being introduced to us. It had been there from time immemorial. Examples are periwinkle (Literina species) from Europe and South Africa. Turbo species in the Pacific sea area have been long known to be very popularly eaten snails in the world. Land snails of the family Helicidae were eaten in1flie Middle East and in Europe edible snails are Helix aspersa ançSmotq. Morocco and Algeria export a lot of snails to the United State of America as food.

In West Africa, land snails, especially those of Genus Achatina are very popular as foods. Snails have been of importance in the diets of the forest dwelling people of West Africa and rank high amongst the meat preference of the Ashanti people of Ghana.

Snails are important source of protein and iron, especially the achatina and Arachachatina species. The iron content of West Africa Giant Snail varies from a locality to another depending on the mineral content of the soil in which the snails are raised.

The fat content of the snail meat is negligible compared with other poultry and other meat.

The percentage of snail of the different parts are as follows: foot edible part 58%, shell 34%, Viscera (contained in the shell) 36% (Wosu, L.O. 2003).

The live weight of individual adult snail is about 200gm. The dressing out percentage of the snail is about 40%. The largest known land snail was found in Sierra Leone in 1976. It was called gee Geromino and weighed 900g and was about 39cm long.

TABLE:

NUTRITION VALUE OF SNAILS COMPARED WITH OTHER FOOD ANIMALS.

From the, above table it is clear that the percentage protein content of a snail is higher than that of any mammalian species. Other areas of the food value of snail show the superiority snail has over other animal or animal products of food origin.

OTHER USES OF SNAIL

shell:

Highly prized by collector

Value high for decoration of homes

Snails are used for making rings

For Cameos production

In Fiji they are used for badges

Used for toys

Used for rubbing walls of houses

B. As fish Baits: Especially the small species

C: The slime can be used as cure for eczema, skin rashes, swells, burns, insect bits in Europe.

D. The ash is very rich in mineral, iron, potassium and other trace element like magnesium, manganese, zinc, copper but no

phosphorus.

Can also be used against hypertension

Abortion

Conjunctivitis

Kidney disease

Anaemia

Asthma and Skin rashes

SOME FOOD AND COMMERCIAL SPECIES OF SNAILS

When we talk of commercial snail production, it is actually the land snails that is applied other than fresh water species since the former are more readily available world wide. The table below shows some of the edible snails species in the world.

TABLE: 10.3

EDIBLE SNAILS

SINO

SNAIL SPECIES

HABIT

COUNTRY AVAILABLE

1

Achatina species.

E.g. - Achatina. The giant snail

-Fulica the garden snail

Land snail (the large size snail

West Africa, East Africa and Nigeria

Arachachatina species e.g Venticosa

Margina (big black) degneri

Land snail (large size)

.

Nigeria, West Africa

Helix (Helisoma) eg Aspersa small grey Pomotia Bug L1y type lucorum (Turkey type) aperta (Borrowing type) nucular

Land snail —

Europe and middle East

Littorina species (periwinkle)

Fresh water/ marine

Europe and South Africa

Strombus species

Marine species

West Indies Morocco and AIga

Ecobama species

Land species

Otala species

ANIMAL HUSBANDRY

Land spies

Morocco and Algria.

Vitrina Species

Land Species

Alpine Meadows ——

Eremma Species

Desert Species .

Egypt

Red Sea Species

Fresh Water

Egypt

Megalatrectus

Land

Australia

Ampullarius

Fresh Water

Nigeria

Limicolaria Species e.g acartensis

Land Snail

BIOLOGY

BIOLOGICAL CLASSIFICATION OF SNAILS

Snails are invertebrate animals of the kingdom. Animalia to which all livestock belong. They are Mulluscs, that is animals with soft body that is covered with hard shell. Further biological classification according to their structure are as.

Phylum Mullusca

Class Univalves

Order Pulmonata

Sub-order Stylommatophera

Family Achatinidae

Genera Achatina

Arachachatina

Gastropods ("belly footed") such as snails and slugs are not nature's ballerinas. When they move, their feet spread out under the body. Snails have coiled or cone-shaped shell. These balance the mass of organs above the foot, like a backpack.

Generally, water-dwelling gastropods have various kinds of gills, and land dwellers have lungs. Gills and lungs are organs of respiration, which enhance gas exchange. Oxygen dissolved in water or inair moves into them, then diffuses into blood. Aerobically respiring cells take up oxygen from the blood. At the same time, the cells give up Carbondioxide, which is moved to the gills or lungs and out from the body.

As most gastropods develop, some body parts grow different rates and become realigned inside the body. By this process, called torsion, the rear of the mantle cavity twist forward and becomes a space into which the head may withdraw in times of danger.

ANATOMY OF THE SNAIL

The anatomy of the snail can be divided into two parts- the shell and the body.

THE SHELL

This is the whorled covering of the snail. The shell consist of various mineral as earlier stated. It is the protective shield into which the snail withdraws at any sign of danger.

The Body: This is the fleshy part in the shell, covered by the mantle, a soft tissue which secrets the shell. The body is made up of the head, the visceral parts and the foot.

REPRODUCTIVE SYSTEM OF SNAIL

The snail is a hermaphrodite. The reproductive organs are known as Ovotesis. This produces sperms and eggs. However, the African Giant snail except Achatina achatina do not self-fertilize themselves. In this process under favourable environmental conditions of feeding, mature snails come together as if to embrace, fertilize one another introducing their developed muscular corpulatory penis into one another vagina and discharge sperms. The sperms fertilize eggs of the receiver-snail. Thereafter they separate to lay clusters of egg, each cluster contains 5-8 egg per lay. As an exception to the rule Achatina achatina self fertilize. Coupling does not proceed reproduction. Egg laying usually takes place in late evening and in the night.

Snails start to reproduce at about 12 months of age. The snail can store sperms received from a coupling partner and use them when it is ready to breed some months later.

Reportedly Arachachatina species take• 3 days interval between copulation and oviposition and incubation period of 38-43 days. Some snails (in all species) carry the developing fertilized egg internally and lay them shortly before they hatch, while some occasionally lay eggs, which are already in the process of hatching. The eggs, glittering glistening and hard when laid remain so unit 24-48 hours before hatching, at the time of which the shells become soft and dull in appearance.

AGE DETERMINATION OF SNAIL

It is generally believed that the age of snail corresponds with or is reflected on the number of rings it has on its shell. Growth in snail is normally and ceases in the third year; hence only two or three annual rings can be seen on the shell. ThiS then means that the rings represents one year (up to the age of three years) after which the aging becomes distorted.

FEEDS AND FEEDING FOR SNAILS

The land snail is omnivorous in nature an.d can survive on almost any form of food. But of all the numerous food savoured by the giant snails. the most popular of plant origin for all ages and sizes of snails in descending order of preference are:

a). Manihot Utilisima (cassava tubers)

b). Carica Papaya (Ripe fruits and leaves)

C). Telfairia Occidentalis (leaves)

d). Amaranthus Spp (African Spinach)

Arachachatina species has preference for the following feed items in the order listed below:

Table:

Cassia fresh leaves

X

KEY: xxxx heighest preference

Xxx highly consumed

Xx moderately consumed

X consumed just a bit

Snail feed on fresh plant material normally but when faced with adverse condition can feed on decaying plant material. The need for balanced diet for snail cannot be over emphasized since this enhances the development of the snail.

COMPOUNDED FEED

A formulated feed can be compound e.g

a. 58.0% cassava peelings

b. 5.8%fresh.meal

c. 6.3% bore meal

d. 0.2% NacI

e. 0.5% Vitamin and mineral mixture.

NIB: Broiler finisher and lavi ration can be used to feed snails.

TEMPERATURE REQUIREMENT

The optimum temperature for excellent performance by snail is between 21°C within the snail house.

REQUIRED RELATIVE HUMIDITY (R.H)

The edible giant snails of West Africa grow actively and reproduce during the rainy season (May-October). They remain quiescent during the dry season (November-April). In the captive population where they are provided with regular water, food and lime (Calcium), snails grow and reproduce throughout the year. The importance of high relative humidity which should not be less than 75% in the snail house cannot be overemphasized

1. Snails in such environment never go into aestivation

2. Any snail that escaped from a moist environment aestivates after 3 days in their hideout and loose weight.

3. On rainy days when relative humidity of the atmosphere was high, the snails emerge from their borrows even in the daytime

PROPER SOIL TYPE FOR SNAIL REARING

Snails depend very much on the soil for food that they can hardly survive or thrive or reproduce effectively without suitable soil. For example, moist, fresh sand is necessary for the deposition, incubation and hatching of eggs of snail, and the young snails eat the soil as one of their first meals in life. The best type of soil for snail rearing is humus soil due to its rich vegetable matter and mineral content and 25% organic matter. It should be moist and cool so that the snails can borrow into the soil during the daylight and move freely on the surface and lay their egg at night.

NEED FOR ADEQUATE SHADE

Snails should be provided with enough shade so as to prevent them from going into aestivation. Aestivation usually affects the reproductive and general performance of snails. The general exposure of snails to sunlight or light generally should not exceed 8 hours per day. Enough air should be allowed into the snail house.

HOW TO SET UP A SNAILERY

Choosing a site:

The following should be considered when choosing a site for a snail farm

a. The site should be in a cool corner of your farm or compound.

b. It has to be shady

c. The soil has to be base. Till the soil before introducing the snails.

d. The soil should be enriched with humus and mineral (Ground bone is good source of mineral)

e. Maintain the soil moisture for adequate humidity.

TYPE OF SNAIL FARMS

There are three basic systems of snail farms. They are:

Intensive systems Semi Intensive Extensive or free range system.

INTENSIVE SYSTEM OF SNAIL FARMING

This is the system where the snails are reared under controlled environmental condition that in many aspects guarantee what they would have had in their natural habitat but modified with the aim of maximizing their productivity.

SEMI INTENSIVE SYSTEM OF SNAIL FARMING

This system is a mixture of both the intensive and the free range system.

It will involve supplementing the feeding of free range snails in an enclosed area with concentrate feeds.

The system could involve using oil palm plantations which provide cool shading environment. In addition to the grasses in between the palms, the farmer may enrich the soil with manure, well ground bone meal, kitchen wastes and stems of banana or plantain trees.

Additionally, this practice could also be carried out in plantain or banana plantation. Snail could also be fattened for a period of time in a colony by feeding and providing them with requisite environment.

EXTENSIVE SYSTEM OF SNAIL FARMING

This involves rearing snails in the wild. The farmer simply goes round to pick them from the wild. It is the cheapest system and least labour intensive.

THREE TYPES OF PENS USED IN INTENSIVE SYSTEM

a. Breeding Pen.

This is for mature snails, which would be allowed to

copulate, lay and incubate their eggs and for the eggs to hatch out the young ones (hatchlings)

b. Rearing Pen

In this pen the hatchlings are gathered and fed to grow up to maturity. From this pen they are transferred to the breeding pen (for those meant for breeding) or to the fattening pen (for those to be sold for food)

c. Growing or fattening Pen

In this pen mature snails from the rearing pen are fed and fattened for sale or for food.

The management in these pens differ because the feeding and other requirements of the snails differ with their sizes, ages and purpose.

SNAIL FARM MANAGEMENT

In housing snails high walls are ideal. The walls should run from foundation to the roof and made of concrete or mud and it should be roofed and ceiled well. The housing is aimed at maintaining conditions of optimum performance by the snails in the way of temperature , relative humidity, and light intensity or darkness. If possible the house should consists of a dark roofing material, cement or concrete walls and the floor of humus soil.

Wooden houses or boxes are also good for snails but above factors should be considered.

Snails house should guarantee space enough to crawl about and their feed should be placed at several points to avoid overcrowding which might result to death or aestivation due to reduced oxygen supply.

HEALTH AND HYGIENE OF SNAIL FARMS

The hygiene here includes all measures that should be taken to observe

the health of the snails and to prevent mortalities. The following should

be guarded against in snail enterprise:

1. Theft

2. Rodents including lizards eat up snails

3. Ants

4. Termites

5. Snakes and hawks

6. Diseases -such as schistomiases, mycobacterium etc

7. Fire

8. Oil spillage

9. Sun rays

MARKETING OF SNAILS

a. Roadside Markets: The snails are displayed as whole snails in baskets, bags or treaded with ropes.

b. Village Markets

At certain seasons of the year Snails are found to be abundant in village markets. Dealers go to the rural areas and buy off the snails.

c. Farm Roads and Fronts of Compounds

d. Export Market

The African giant snails are becoming increasingly popular in Europe and America. Europe, France, Germany, UK, Belgium, Denmark are becoming big markets for snail.

PROCESSING AND WASHING OF SNAILS

Processing:

a. By crushing the snail shell with a hard object so as to remove the entire body mass.

b. By putting the snail in boiling water

WASHING

a. Use of alum

b. Use of lime

c. Water: After above wash the snail twice with clean water for th ,h cleansing.

COOKING AND CANNING

After processing and washing, the snail is ready for cooking or canning

(for local or export market).

CHILLING OF SNAILS

Snails may be transported chilled both for export and locally. It is important to maintain the chilled state.

STEPS TOWARDS SETTING UP A SNAIL FARM

1. Choose the spot for the farm in a quiet and shaded section of your compound or farm.

2. Make a good feasibility study of the operation before you start, no matter how small or large the farm.

3. Build appropriate house or structures for the size and type of the enterprise you want. Do not forget to construct a gutter around the snail house.

4. Ensure that there will be adequate availability of the feed and water for the snail as needed.

5. There is need for adequate humidity and soil dampness in the farm for the snail all the time including the dry season.

6. Ensure the right type of soil is placed in the house at the start.

7. Seed the farm with healthy and the right of snails ensuring there is no crack.

8. Ensure p:r routine management of the farm all the time.

9. Be ready t6 market your product at the right time

Chapter Eleven

AGRO-FEASIBILITY STUDY

Agro-feasibility study involves the putting down in a written form of what a farm enterprise should be, what it could cost, what profit it could give, the life span of the infrastructure, the equipment and whether the enterprise will be viable or no.

In modern agricultural enterprises Agro-feasibility has become a must since modern day agriculture has become as serious and capital intensive, considering all factors of production like any other business endeavor.

For any agro-business to take off and succeed there must be a working document which the farmer must follow Sacrosanctly if the venture must not abort.

Additionally, without a detailed and professionalized study, it is virtually impossible for any financial institution to finance an agricultural business. Most agriculture business have collapsed and are still collapsing till date because there have been no working document from which a good foundation would have been Iaid.It is like errecting a structure without the appropriate architectural and engineering drawings and adhering to same rigidly. The result will be an imminent collapse of such a structure. The same thing applies to agro- business.

Accordingly, I thought it wise to include this all important subject in this write up for the benefit of my readers. A copy of a model feasibility study has been included in this write up for their benefit.

AGRO-FEASIBILITY STUDY FOR CIVIL THRUST

CO-OPERATIVE SOCIETY NIGERIA LIMITED

A FEASIBILITY STUDY BASED ON POULTRY

PRODUCTION PROJECT IN OBIO/AKPOR LOCAL

GOVERNMENT AREA OF RIVES STATE

BY

JUMEZ NIGERIA,

322 ABA ROAD,

HENRY OJO

RUMUOKWURUSI, PORT HARCOURT,

PHONE: 08035500314

E-MAIL: zurique2000yahoo.corn

APRIL 2004

INTRODUCTION

CIVIL THRUST COOPIRATIVE SOCIETY NIGERIA

LIMITED has a parcel of land of over four hectares in Obio/Alcpor Local Government Area of Rivers State of Nigeria, some five kilometers from the city of Port Harcourt.

A specialized system is to be adopted for this project and will comprise mainly eggs production and broilers or cockerels when necessary. There will also be a vegetable production unit that will make use of the manure from the farm.

The project will involve the following:

a) 1000 layer unit.

b) Vegetable production unit

The project will provide eggs and quality vegetables all

the year round for civil servants in the state at a very reasonable price. It is also hoped that the farm will be supplying eggs and vegetables to shops and groceries in the city oF Port 1-larcourt. V

MARKET

The farm is located in a strategic position and very much accessible by road from Port Harcourt the state capital.

it is expected that arrangement will be made with cateri.ng outfits, supermarkets, at both state and Federal Secretariats as outlets for the products from the farm.

It is also expected that customers would even book in advance for the products of the farm. The eggs would be sold m crates and also parceled in dozens. Additional arrangements would be mad.e with market women, schools, hotels and universities for supply of the farm products.

Spent layers (old layers) would be sold as culls after being processed, packaged and frozen for sales to minimize the cost of feeds. However the sale of live spent layers cannot be ruled out as most Nigerian housewives prefer to slaughter their birds themselves.

Vegetable could be sold in the open market and arrangements would be made to supply to sOme catering outfits. The manure for growing the vegetable will be provided. from the poultry itself. Extra poultry manure will be sold to farmers and flower

gardeners. No extra labour will be needed in the vegetable farm as the existing attendants in the poultry unit will be deployed from time to time to the vegetable farm.

SECTION I

THE POULTRY SECTION

1.1 THE LAYER UNIT

This unit is for table egg production in commercial quality under an intensive system of management.

CONSiDERATION: Deep litter system i.e. a system where birds (layers) are kept on a deep litter throughout their rearing and laying periods as against the battefy cage system where the birds are transferred to the cages at the comment of their laying period.

1.2 DESIGN CONCEPT

It is proposed that 1100 day-old chicks (pullets) shall be purchased for the project and reared up to point of lay. The layer house shall be able to contain up to 5000 layers for the future expansion of the project. The structure will be approximately 38m x 1 im. A brooder house shall be considered in the future measuring 54 x 7.2m.

The house will be approximately 2.5rn high at the eaves and 3.6m high at the ridge.

The upper I .5m of the sides will consist of a series of opening windows to allow for light and ventilation. The roof having a pitch of about 20° will be clad in corrugated alurninim sheeting or similar material and there will be a I rn overhanging of the caves, 3.6rn at the ridge, 0.45rn dccl) foundation, and 0.7m thick, walls 0.23m thick and im high.

An open cowling will be provided over the centre ridge to allow the escape of hot and humid air.

Conical hanging feeders will be used as well as movable drinkers. Empty drums will be cut to sizes and used as laying nests for the birds. Fresh litters will be used daily in the drums to ensure clean eggs. Fresh water and adequate feeds shall be given to the birds to ensure maximum performance.

1.3 LAYING

At point of lay (18wks) the birds shall be taken to the other side of the pen which will serve as the layer house and feed changed from grower mash to layer mash. The layers will continue to lay for about 52 weeks and it is estimated that egg production will not be less than 70% daily. Mortality is estimated at 10% before onset of lay and another 10% at the end of lay. Restocking will be every 9 months. When in full production there will be an output of nothing less than 770 egg daily from the 1100 layers.

The spent hens will be sold and again when in full production, about 880 spent hens will be sold annually assuming a 10% mortality during laying.

1.4 FEEDING

Feed requirements are based on the following schedule:

0-8WKS

4.5kg/ 100 chicks per day of Chicks Mash

9WKS-18WKS

.0kg per 100 pullets per day of Growers Mash

19WKS — 7OWKS

12.5kg per 100 layers per day of Layers Mash

Feeds would be purchased from reputable feed millers and enough quantities should always be in stock to prevent any feed scarcity which might disrupt the production schedule.

1.5 MARKETING

The intention is that eggs will be packed in the farm and sold to traders who will come to the farm to collect egg. Efforts will also be macic to send egg to certain centres in the state capital.

Good prices will be achieved at N400 per crate of 30 egg. Spent hens (alter laying) will be sold and it is anticipated that the price of N600 per bird will be achieved.

1. .6 STAFF REQUIREMENT

The project will require an attendant under the management of a supervisor whose minimum qualification should be HND/BSC in Animal Science. This attendant will also be able to take care of the vegetable section.

1. .7 THE COOPERATIVE SOCIETY

TI-IE CIVIL THRUST COOPERATIVE SOCIETY NIGERIA LIMITED is a Limited Liability Cooperative Society and is incorporated in Nigeria and' registered in the Companies Registry in Abuja as No The shareholders are members of the cooperative society who shall buy off the shares of the cooperative. Interested members of the public shall be considered.

The authorized share capital of the cooperative society shall be 1 1,000,000 made up of share that will be determined by the cooperative society.

2.5 RECURRENT COSTS ASSUMPTION

Costs of labour as indicated in the recurrent costs table shall be the same for the vegetable section. Day old chicks (pullets) will cost N220 each while feeds will go as follows:

Chicks mash Growers mash Layers mash

Ni 150 per 25kg bag

N900 per 25kg bag

N1050 per 25kg bag

Veterinary care at N5,000 monthly Power generation and main tens nec

3.2 PROJECT VIABILITY

From the project analysis and cash flow projection it is shown in clear terms that the project is very feasible and viable.

3.3 LOAN REQUIREMENT

Total working capital investment in the first year is close to NI 1,000,000. Part of the amount will be expected to come from loan and the rest to be footed by members of the cooperative society.

3.4 FUND UTILIZATION

The loan and other funds available to the cooperative society would be utilized for the development of this project via the acquiring of land, construction of poultry houses, equipment and for overhead costs.

3.5 WORK ALREADY DONE

No work has so far been done concerning the project. It is however expected that soonest the project will kick off.

3.6 TAXATION

Tax payable is 2% of the total revenue for each year and then is assumed to be deducted from source.

3.7 LOAN REPAYMENT AND INTEREST RATE

An interest rate of 17% is envisaged and the loan will be paid over a period of 4 years with the first one year being year of grace.

BIBLIOGRAPHY

G. Williamson and J.A.Payne (1978): An Introduction to Animal

Husbandry in the Tropics.

Longman Group Limited,

London

M.F. Komolafe et al(1 979) Agricultural Science for West African Schools and Colleges. University Press Limited, Ibadan

J.A. Oluyemi and F.A. Roberts(1 979): Poultry Production in warm wet Tropics. The Macmillan Press

Ltd London and Basingstoke

L.O. Wosu (2003): Commercial Snail Farming in West Africa. A Guide . AP Express Publishers limited, Nsukka, Nigeria.

S. Pichi and W.J.. Peters (1985): Possibilities of using the grass cutter for meat production in Africa. Institute of Animal Husbandry and Genetics, University of Cottingen.

Ojuka Partner D: The effects of feeding regimes and floor space on the performance of young grasscutters.

Rivers State University of Science and Technology, Port Harcourt (unpublished)

(1991): Rabbits Production in the Tropic. The Tropical Agricultural Series Macmillan Education

Ltd. London and Basngstoke

Nancy Roper (1933): Pocket Medical Dictionary. The English Language Book Society

and Chirr-h Livingstone

ABOUT THE AUTHOR

Henry Ojo is one of the end time teachers and Gospel preachers. He has pastored and planted so many churches. He had personal encounter with the Lord Jesus Christ in 1979, also laboured in the seminary as lecturer and grew to the position of provost. He is presently the organizer of minister conference in Christ Apostolic Church. He is happily married to Ruth Olufunsho and the marriage is blessed with four children: Emmanuel Temitope, Christana Olamide, Jonathan Oluwatobi and Victoria Abosede.